I0093835

Francis Polkinghorne Pascoe

Notices of new or little known genera and species of coleoptera

Francis Polkinghorne Pascoe

Notices of new or little known genera and species of coleoptera

ISBN/EAN: 9783337221737

Printed in Europe, USA, Canada, Australia, Japan

Cover: Foto ©Andreas Hilbeck / pixelio.de

More available books at **www.hansebooks.com**

NOTICES

OF

NEW OR LITTLE-KNOWN

GENERA AND SPECIES OF COLEOPTERA.

BY

FRANCIS P. PASCOE, F.L.S., ETC.

[*From the* JOURNAL OF ENTOMOLOGY *for April.*]

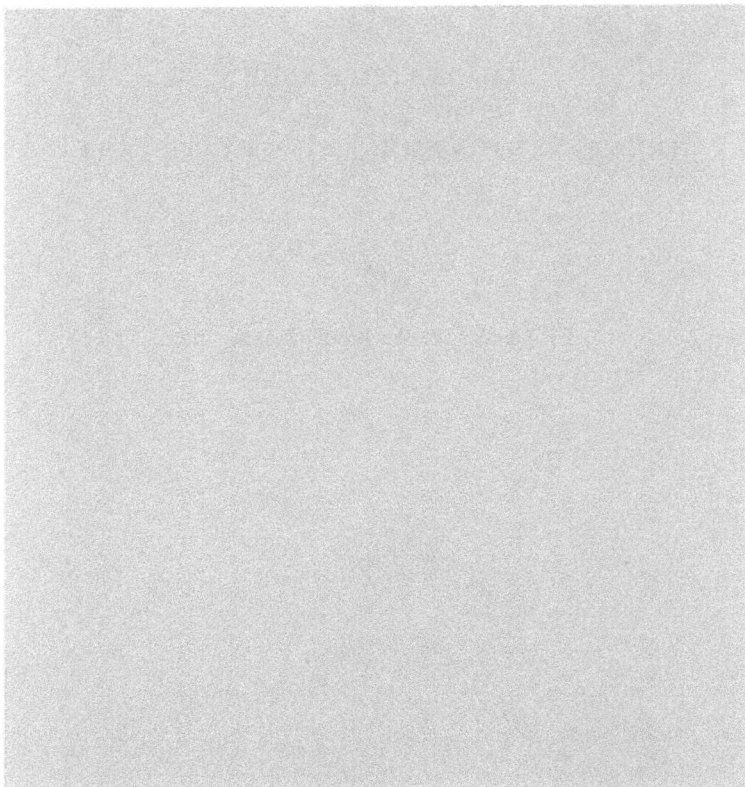

Notices of new or little-known Genera and Species of Coleoptera.

By FRANCIS P. PASCOE, F.L.S., &c.

PART I.

IT is difficult to form any adequate idea of the number of new forms, to say nothing of new species of insects, which exist in, or are being constantly added to, our cabinets*. Many of these are almost hope-

---

* Mr. S. Stevens has just favoured me with the sight of a collection of Coleoptera (perhaps about a thousand species) made by Mr. Squire at Petropolis (a sort of Brazilian Cintra, and a short day's journey from Rio), and although the district has been repeatedly worked, and Mr. Squire was there scarcely two months, yet the result of his visit has been the discovery of a vast number of novelties and some new forms of a very interesting character.

lessly entombed in private as well as in public collections, or have long been accumulating in my own. To record the most remarkable, and such, at the same time, as can *easily* be recognized by figures and descriptions, if confined to a private collection, is one of the objects of this Journal, and the following is the first of a series of papers which will be devoted to the Coleoptera. As it will be impossible to follow any systematic plan beyond the limits of each paper, a classified list will be given hereafter to diminish this inconvenience.

It must not be forgotten that many of the insects to be described will be either uniques, or, belonging to others, cannot therefore be mutilated by dissection ; but as every new genus will be figured, it is hoped that the absence of the usual analyses will not create any difficulty. *Practically*, we are satisfied with referring species to their genera from their external resemblances; but although it is very often quite impossible to ascertain the affinities of an insect without dissection, there is the danger of attaching too great importance to organs whose characters cannot always be determined satisfactorily, and which, moreover, because they occur in one species, are sometimes erroneously assumed to be present in others. Indeed, it may be doubted if even individual species are so invariable as to justify the minute descriptions of many naturalists.

While believing in the existence of genera quite as much as in the existence of species, is it satisfactorily established that they can always be distinguished by technical characters, such as we are in the habit of employing? In all large genera, I believe, it will be eventually found that they possess no one character in common that is not also possessed by the group or family to which they respectively belong, and hence it is quite natural that the limits of such genera cannot be very strongly defined. This is especially the case in the Longicorn families, which with endless differences in habit agree in a certain similarity of details, so that the generic characters often become mere questions of degree,—while, on the other hand, many Heteromera alike in habit are found to vary remarkably in structure, and in fact to belong to very different groups than those in which their general appearance would seem to place them.

These and other points of the same kind will be often exemplified in the course of these ‘ Notices ; ’ but in considering the difficulties which beset all attempts at a satisfactory limitation and arrangement of species into families, genera, &c., it will be as well to bear in mind the remark of our great naturalist,—“ Nature is less of a systematist than Man.”

OMOPHRON [Carabidæ].

Latreille, Hist. Nat. des Ins. viii. p. 278.

Omophron Brettinghamæ.

O. ovato-rotundatum, nitidum, testaceum; capite, prothoracis disco, elytrisque (marginibus exceptis) viridi-æneis.

Hab. India (Dacca).

Shortly ovate or nearly orbicular, moderately convex, very smooth and
shining; head sparingly punctured, brassy-green; labrum, epistome, and
small triangular spot above the latter brownish-testaceous; prothorax
finely and remotely punctured, and with the elytra rich brassy-green,
bordered externally with testaceous,—the border much wider on the
latter, which are also very finely punctured in rather distant rows;
eyes and tips of the mandibles dark brown; antennæ, palpi, and legs
pale testaceous; body beneath with the sterna pitchy, the abdomen
deep testaceous.   Length 2 lines.

For the possession of this and many other Coleoptera from the
same locality, I am indebted to Dr. Ernest Adams of University
College, the author (*inter alia*) of some exceedingly interesting
and learned papers on the "Vernacular names of Insects," in the
Transactions of the Philological Society, who received them from
India*, where they were collected by Mrs. Brettingham (to whom I
have dedicated the exquisite little *Omophron* just described) in the
compound attached to the quarters of Charles Brettingham, Esq., in
medical charge of the Kamroop Regiment of Native Infantry sta-
tioned at Dacca.   They comprised above seventy species, belonging
to nearly as many genera.   Of these there were only six or seven
that were not represented in Europe, viz. *Adoretus, Heteronychus,
Anisotelus, Macratria,* a Nitidulid, and two, or perhaps three, obscure
Heteromerous genera, which I have not been able to refer to any
hitherto published.   Except that there were very few Staphylinidæ,
they were mostly such forms as would be found in this country in
the *débris* of a flood; and it is, therefore, most likely that they
were collected in the rainy season.   Dacca is nearly in the latitude
of Calcutta, lies very low, and as it is subject to inundations from
the Ganges, it is probable that it affords a larger proportion of
European forms than would have been the case in a drier or more
elevated district.   So little is *really* known of the Entomology

---

* Upwards of a thousand specimens, some nearly an inch long, although generally much smaller, enclosed in two large-sized pill-boxes, were transmitted by
post in the ordinary way in a single letter.   A wine cork hollowed out in the
middle, and a little trimmed at the sides, would be an excellent, and at all times
available substitute for a box.

of India, that it would be premature to speculate as to its character; but although in its animal productions there is a remarkable amount of Northern temperate forms, they could never have
been expected to predominate to such an extent in any one group,
as it now appears they do, judging, however, solely from the facts
here stated*. There is one point in connexion with this subject
which can only be just alluded to here, viz. the apparent tendency of
animals to migrate to the south, and not the reverse, or very slightly
so†. The idea first occurred to me in noticing the very few Australian forms of Coleoptera occurring in Mr. Wallace's Indian Island
collections contrasted with the number of these Indian forms in
Australia, especially its northern parts;—compare also Natal with
the Cape, or the United States with Mexico or Cuba; notice Upper
Egypt, Arabia, Persia, &c., just receiving a tropical form here and
there.

CASNONIA [Carabidæ].

Latreille, Icon. de Coleop. d'Europe, ed. 1. p. 77.

Casnonia aliena.

C. picea; capite infra oculos profunde lunato-impresso; prothorace
capite breviore, postice transverse corrugato; elytris leviter striatis,
singulo macula flavescenti apicem versus ornato.

Hab. Australia (Moreton Bay).

Pitchy, shining; head rather broad, a deep semicircular impression
between the eyes and the epistome; prothorax not so long as the head,
somewhat elliptical, smooth anteriorly, but with delicate transverse
folds behind; elytra about the length of the head and thorax together,
and three times the width of the latter at its base, faintly striated with
an oblong yellowish patch towards the apex of each; antennæ dark
brown; legs pitchy, femora at the base and tibiæ in the middle (nearly
obsolete, however, in the anterior pair) testaceous-yellow; body beneath
pitchy-brown, shining. Length 5 lines.

Although the genus *Casnonia* is found in India as well as in
South America, this is, I believe, the only species yet detected in

* While this sheet was passing through the press, Dr. Adams received another
collection from Dacca, evidently made in a more favourable season; still, although
there is an addition of many tropical genera, European vastly predominate; and
it is worth notice, that nearly all these tropical genera have a very wide range, as
for example, *Anthia, Chrysochroa, Protætia, Xylonychus, Cerosterna, Olenocamptus,
Glenea, Apomecyna, Batocera, Xylorhiza,* &c., all of which are represented by the
commonest species.

† The same tendency has been noticed in plants, so far as those of America
and Australia are concerned.

Australia, and so far it appears to be absent from the Indian Islands. The present insect is rather larger, with a shorter thorax, than any *Casnoniæ* that I am acquainted with.

Sostea [Parnidæ].

Head small, completely retractile within the thorax. Antennæ received, in repose, in a cavity beneath the eye; 11-jointed, the first very large, laminate, the second dilated inferiorly, the remainder forming a compact flabellate mass. Eyes rounded, entire. Mandibles bidentate at the apex. Maxillæ with lobes, short and broad. Maxillary palpi short, the last joint very large, cylindrical; the labial claviform. Mentum transverse, lobed in front. Labium dilated anteriorly. Prothorax transverse, convex, nearly semicircular. Elytra ovato-triangular, very convex, gibbous towards the base. Legs slender, coxæ distant, tarsi short. Prosternum received in a notch of the mesosternum.

These characters are drawn up exclusively from *S. Westwoodii*; but the other species so completely resemble it, that there can be no doubt as to their generic identity. In all, the elytra have nine rows of punctures on each. The structure of the antennæ will be better comprehended by the figure* than by any description, but owing to their minuteness, I was unable to detach completely the large basal joint; when in repose it appears to act as a valve, closing in entirely the rest of this organ: all the joints composing the flabellate mass are what may be called boat-shaped, except the last, each being received at its base, and for the greater part of its length, in the concavity of the preceding one,—the first, however, being so much more dilated as so far to enclose the succeeding or fourth joint, that it is only visible at its free extremity; and unless this is attended to, the antennæ will appear to be composed of ten joints only.

I have dedicated the first species to J. O. Westwood, Esq., M.A., F.L.S., &c., and have adopted his views respecting the position of the genus, of which, indeed, there can be no doubt.

Sostea Westwoodii.   (Pl. II. fig. 6.)

S. ovata, fusco-atra, nitida, longe setosa; scutello triangulari; pedibus rufescentibus.
Hab. Borneo (Sarawak).

Ovate, shining brownish-black, covered with scattered long black setose hairs; prothorax sparingly punctured; scutellum triangular; elytra strongly punctured; legs brownish-ferruginous. Length 2 lines.

---

* This plate was a first attempt at drawing on stone.

Sostea carbonaria.

S. suboblongo-ovata, atra, nitida, breviter setosa; scutello oblongo: tarsis rufescentibus.

Hab. Moluccas (Batchian).

Ovate, a little inclining to oblong, shining black, with short scattered black setose hairs; scutellum oblong; tarsi ferruginous-red. Length 1¼ line.

Smaller than the last, with which it agrees in shape (except that it is a little longer proportionally), punctuation, &c.; but it is at once distinguished by its oblong-ovate scutellum.

Sostea æneipennis.

S. ovata, nigra, nitida, longe setosa: scutello triangulari; elytris æneis; pedibus rufo-ferrugineis.

Hab. Borneo (Sarawak).

Ovate, shining bluish-black, with long setose hairs; prothorax rather sparingly punctured; scutellum triangular; elytra brassy; legs reddish-ferruginous. Length 2¼ lines.

Sostea cyanoptera.

S. ovata, atra, nitida, setosa; scutello triangulari: elytris læte cyaneis; pedibus ferrugineis.

Hab. Borneo (Sarawak).

Ovate, shining black, with moderately long setose hairs; prothorax finely punctured; scutellum triangular; elytra rich ultramarine blue; antennæ pale testaceous; legs ferruginous. Length 2 lines.

Sostea secuta.

S. ovata, fusca, subnitida, setosa; prothorace griseo-pubescenti; elytris obscure cyaneis; pedibus ferrugineis.

Hab. Borneo (Sarawak).

Ovate, dark brown slightly inclining to reddish, with long setose hairs; prothorax covered with a short thick greyish pile; scutellum triangular; elytra deep indigo-blue, shining but slightly, with a pale, thin pubescence; legs ferruginous. Length 1¼ line.

Resembles the last, but is smaller, with a very decided pubescence, which is almost absent in the rest of the genus.

Sostea elmoides.

S. breviter ovata, fusca, longe setosa, fortiter punctata; scutello sub-cordato; pedibus rufo-ferrugineis.

Hab. Borneo (Sarawak).

Shortly ovate, dark brown, with long setose hairs, strongly punctured above; scutellum subcordate; legs reddish-ferruginous. Length 1¼ line.

Broader and more rounded at the apex of the elytra than any of its congeners, and proportionably shorter.

Byrsax [Colydiidæ].

Head small, vertical, hidden above by the prothorax. Eyes large, rounded, partially divided by the cheek. Antennæ retractile, short, gradually increasing upwards; the first joint rather slender, the second shortest, the third and fourth longest and equal, the fifth triangular, the sixth to the tenth transverse, the eleventh shortly ovoid. Labrum and epistome very small. Palpi short, linear, the terminal joint ovate. [Mentum transverse; labium oblong, entire, as seen *in situ*.] Prothorax very transverse, gibbous in the middle, bicornuted anteriorly, the margins dilated and crenulate. Elytra short, very convex, tuberculate, with broad crenulate margins. Legs of moderate size; tarsi with the first three joints very short, equal, with fine hairs beneath. Prosternum strongly compressed. Mesosternum toothed.

In habit this genus closely resembles the *Diaperis horrida*, Ol., with which Mr. Walker's *Asida horrida* is probably identical. Its real affinity, however, if we are to be guided by the tetramerous tarsi, is with *Endophlœus*, *Pristoderus*, and some other little-known and even undescribed forms among the Colydiidæ, but differing from all in its head being perfectly hidden by the prothorax when viewed from above*.

Byrsax cœnosus.   (Pl. III. fig. 7.)

B. rotundatus, pellicula fusco-murina indutus, infra piceus; antennis palpisque brunneis.
Hab. Singapore.

Nearly orbicular, very convex, dark brown, covered with a thin yellowish-brown pellicle, which readily peels off; prothorax with two short porrect horns in front; scutellum small, triangular; elytra each with three tubercles placed near the suture, the two anterior much the largest; body beneath pitchy; antennæ and palpi light brown. Length 4 lines.

The figures represent the head as seen from below, and the intermediate tarsus.

Sphæromorphus [Scarabeidæ].

Germar, Zeitschr. für d. Entom. iv. p. 111.

Sphæromorphus acromialis.

S. convexus, fusco-piceus; prothorace antice elevato, basi inæquali; elytris suboblongis, elongato-punctatis, humeris elevatis bituberosis.
Hab. Singapore.

---

* The male (which I have only just noticed in the British Museum) has two long erect horns on the head. The same collection contains a second species of this genus, also from Singapore.

Convex, dark pitchy-brown; head rather broad and a little flattened in front, finely punctured; prothorax very transverse, with numerous areolated punctures, tumid anteriorly behind the head, the sides and disc somewhat concave, the base with two round prominences on each side and a transverse raised line behind them, the anterior angles short, obtuse; scutellum triangular, lying in a hollow between the elytra; elytra shining, slightly oblong, covered with irregular elongate punctures, elevated at the base, the shoulder with two tuberous prominences; antennæ, palpi, body beneath, and all parts of the legs not exposed when the animal is rolled up, pale ferruginous. Length 2 lines.

Sphæromorphus Wallacei.

S. subdepressus, nigro-piceus; prothorace æquo; elytris rotundatis, basi paullo convexis.
Hab. Borneo (Sarawak).

Subdepressed, dark pitchy inclining to black; head slightly convex, finely punctured; prothorax smooth, even, with minute areolated punctures, its anterior angles rounded; scutellum very large, triangular; elytra with a nearly rounded outline, the base towards the shoulders very slightly convex, covered with delicate elongate punctures; antennæ, palpi, body beneath, and legs, where not exposed when the animal is rolled up, ferruginous. Length 1½ line.

The occurrence of a genus so purely American as *Sphæromorphus* in Borneo may well excite surprise, as, *à priori*, it might have been supposed, if any of that group occurred at all in the Indian Archipelago, it would have been either a new form, or the Madagascar *Synarmostes*. I cannot, however, find, from dissection of *S. acromialis*, any variation of character sufficiently marked to warrant its separation from *Sphæromorphus*. Dedicated to Mr. A. Wallace, to whose researches in the Indian Archipelago we owe so much.

Idgia [Telephoridæ].

Laporte de Castelnau, in Silberm. Rev. Ent. iv. p. 27.

Idgia flavirostris.

I. viridis; capitis fronte nigra; rostro, prothorace, femoribusque flavo-testaceis.
Hab. North China.

Elongate, deep green, scarcely shining; head thinly punctured, a deep ʌ-shaped impression between the eyes, front to just below the eyes black, rest of the head and palpi yellow; prothorax yellow, subquadrate, a little broader than the head, its sides towards the base somewhat concave with a longitudinal impression in the centre; scutellum obtuse behind; elytra deep green, narrow (from contraction

in drying appearing acuminate at the apex), very minutely punctured
with small granular points principally on the basal half, and sparingly
covered with short stiffish hairs (invisible except under the lens);
antennæ about half the length of the body, the four basal joints yellow,
the remainder dark brown; legs slender, coxæ and femora testaceous-
yellow, tibiæ and tarsi brown; body beneath black, breast and sides of
the abdomen pale yellow.   Length 6 lines.

Dascyllus [Dascyllidæ].

Latreille, Précis de Carac. gén. des Ins. p. 43.

Dascyllus congruus.

D. elliptico-ovatus, fusco-piceus, griseo-pilosus; antennarum articulis
subcylindraceis.

Hab. North China.

Ovate-elliptical, pitchy-brown, everywhere covered with short, coarse
greyish hairs; scutellum broadly cordate; joints of the antennæ nearly
cylindrical (particularly the last seven).   Length 6 lines.

Closely allied to the European *D. cervinus*, but larger and more
robust, the thorax a little longer, the scutellum less transverse, and
the joints of the antennæ more cylindrical, or rather less contracted
at the base.

Cylidrus [Cleridæ].

Latreille, Fam. Nat. p. 354.

Cylidrus centralis.

C. piceus; plaga magna fulva communi medio elytrorum; pedibus quatuor posticis testaceo variis.

Hab. Moreton Bay.

Pitchy-brown, very glossy; head and prothorax finely punctured;
elytra minutely punctured in rows, a large, nearly median fulvous-yellow
patch common to both; palpi and four or five basal joints of the antennæ
fulvous; middle and posterior legs, especially the latter, testaceous,
slightly varied with brown.   Length 3 lines.

C. nigrinus, from Tasmania, is, I believe, the only species of this
widely diffused genus hitherto described from the Australian pro-
vince.

Cylidrus alcyoneus.

C. cyaneus; capite chalybeo-atro; femoribus testaceis; antennis nigris,
basi palpisque fulvis.

Hab. New Guinea (Dorey).

Rather narrower than *C. cyaneus*, Fab., and very glossy: head bluish-black, finely punctured; prothorax metallic green, sometimes blue, slightly corrugated at the side, coarsely punctured at the anterior margin; elytra rich indigo-blue, with a few scattered pale yellowish hairs; antennæ black, the four basal joints and palpi fulvous; legs testaceous, tibiæ and tarsi varied with brown; abdomen, and sometimes the metasternum, brownish-testaceous.   Length 5 lines.

ELEALE [Cleridæ].

Newman, The Entom. p. 36.

Eleale sellata.

E. chalybeo-viridis; prothorace pedibusque nigro-æneis; elytris angustis, singula plaga magna elongata, antennisque flavis.

Hab. Moreton Bay.

Rather narrow and subdepressed, covered with long black setose hairs; head with numerous shallow punctures, dark bluish-green; prothorax transversely corrugated, brassy-black; scutellum covered with white hairs; elytra rather elongate, a little contracted posteriorly, closely and deeply punctured in nearly regular lines at the base, more dispersed towards the apex, which has a slight fringe of greyish hairs, dark steel-blue, each with a long fulvous patch extending from the shoulder to about two-thirds of its length, but not meeting at the suture; legs brassy-black; antennæ yellow; eyes brown.   Length 4 lines.

Eleale lepida.

E. aureo-viridis, modice elongata; elytris purpureo-atris, fasciis duabus, scutelloque fulvis.

Hab. Moreton Bay.

Moderately elongate; head and prothorax thickly punctured, deep golden-green; elytra slightly contracted in the middle, coarsely punctured, dark purple-black,—a broad band nearly in the middle, another at the apex, and the scutellum fulvous-yellow; legs brassy-black, the tibiæ more or less fulvous; eyes black; antennæ yellow; body beneath coppery, with long greyish hairs.   Length 5 lines.

Eleale simulans.

E. aureo-viridis, breviuscula; elytris purpureo-atris fasciis duabus fulvis; scutello concolore.

Hab. Moreton Bay.

Closely resembles the last, but is smaller and proportionably shorter, the sides of the elytra parallel, the scutellum black, the eyes dark blue, the head and legs with a decided bluish tint, &c.   Length 3½ lines.

SCROBIGER [Cleridæ].

Spinola, Monog. de Clérites, i. p. 230.

Scrobiger albocinctus.

S. ater; prothorace subtilissime punctato; elytris fasciis duabus albis, una subobsoleta, altera, pone medio, obliqua.

Hab. Moreton Bay.

Nearly allied to *S. idoneus,* Newm., but the eyes are smaller and less prominent, the prothorax more finely punctured, the anterior band on the elytra nearly obsolete and more median, and the posterior *directly* oblique, not curved.   Length 5 lines.

CORMODES [Cleridæ].

Head rather short, broad in front.   Eyes ovate, vertical, scarcely emarginate. Antennæ as long as the thorax, arising laterally in front of the eyes, 11-jointed, the first largest, the second shorter than the third, the last three forming a slender pointed club. Palpi with the terminal joint of the labial securiform, of the maxillary cylindrical.   Labrum small, hairy. Prothorax subdepressed, rounded in front and at the sides, contracted posteriorly,—the pronotum confounded with the parapleura. Scutellum transverse. Elytra depressed, narrowed at the base, gradually expanding at the sides, with a strongly marked carina at the shoulder, but no humeral angle. Wings none. Legs stout, femora clavate, tibiæ and tarsi short, the first tarsal joint nearly covered above by the second; claws simple. Abdomen of five segments.

Although very dissimilar in habit to the Cleridæ in general, there is no doubt that this genus is closely allied to *Natalis.*   It is, I believe, the only one of its family without wings,—a condition due, as Mr. Darwin tells us, in reference to other insular apterous Coleoptera, to " the action of natural selection, but combined probably with disuse," and therefore it would not, perhaps, be very difficult for the advocates of his theory to suppose *Cormodes* a descendant of *Natalis,* to which it certainly bears a very peculiar resemblance. The absence of a real humeral angle, but its simulation by an elevated and narrow carina (absent in all other Cleridæ), and the, in other respects, well-developed elytra, do not appear to lead to the conclusion of the gradual reduction of the wings which such an explanation implies, because corresponding with this presumed reduction we have an unaccountable and apparently unnecessary increase of the elytra, combined, however, with the absorption of the humeral angle. I possess a Longicorn, closely allied to Mr. Wollaston's oceanic genus *Deucalion,* also without humeral angles, but having perfect, although excessively small, wings, and of course entirely useless for the purpose of flight : but in this case the wings might at any time disappear from physical causes alone, just as we find certain species of Hemiptera becoming apterous in cold localities or in very cold seasons. In these and other instances of abnormal variation, which in almost every case seem to have some speciality of their own, we look in vain for the "advantage" which is supposed to have been acquired in the "struggle for life." An insect so suggestive of Mr. Darwin's theory should appropriately bear his name.

Cormodes Darwinii. (Pl. II. fig. 8.)

C. testaceo-brunneus, fere piceus, hirtis sparsis indutus; prothoracis medio sulcato; elytris pallidioribus, seriatim punctatis.
Hab. Lord Howe's Island, South Pacific.

Pale testaceous-brown inclining to pitchy, particularly on the prothorax and base of the elytra, and everywhere but very sparingly covered with loose greyish hairs; head punctured in front; prothorax with a short deep longitudinal impression in the centre; elytra rather wider than the base of the prothorax, with a strong basal carina, which gradually disappears at rather beyond half their length, the shoulder with another strong carina which is continued nearly to the apex, the side beneath the outer carina bent inwards at the shoulder, coarsely and regularly punctured, the punctures becoming smaller posteriorly; mandibles pitchy; eyes brown. Length 7 lines. British Museum.

Aulicus [Cleridæ].

Spinola, Rev. Zool. 1841, p. 74.

Aulicus viridissimus.

A. subangustus, chalybeo-viridis, nitidus; antennis fusco-luteis; pedibus atro-cyaneis, gracillimis.
Hab. Australia (Sydney).

Rather narrow, dark chalybeate green, shining, with sparse, long, black, setose hairs; head and prothorax coarsely punctured, the latter with a deep transverse groove anteriorly, and a longitudinal one in the centre; elytra about two and a half times longer than broad, thickly and coarsely punctured in rows; legs (especially the posterior pair) slender, dark blue; body beneath shining greenish-blue. Length 3 lines.

Aulicus lemoides.

A. latior, aureo-viridis, nitidus; capite prothoraceque cupreis; antennis flavis; pedibus piceis, femoribus basi apiceque testaceis.
Hab. Australia (Moreton Bay and Sydney).

Rather broad, golden-green, shining, with numerous pale greyish setose hairs; head and prothorax rich copper-red, sparingly and rather less coarsely punctured, the latter with the transverse impression nearer the anterior border, and with the longitudinal one rather less deep than in the last; elytra only twice as long as broad, coarsely punctured in rows; antennæ, palpi, mouth, and throat pale yellow; legs pitchy, stout, femora at the base and apex (or legs altogether) testaceous; body beneath green, more or less covered with greyish hairs. Length 3 lines.

Aulicus instabilis, Newm., the type of the genus, is such a variable insect, that it is quite possible this may be but one of its protean forms; nevertheless, besides its smaller size, it is more convex, the prothorax narrower and less depressed, its greatest breadth being behind the middle, and the posterior and anterior margins being nearly equal; its head is also shorter, the eyes proportionally larger, and the antennæ longer; moreover I have never seen any specimen of *A. instabilis* approaching this in colour.

ALLELIDEA [Cleridæ].

Waterhouse, Trans. Ent. Soc. vol. ii. p. 193.

Allelidea brevipennis. (Pl. II. fig. 9.)

A. elongata, atra, nitida; elytris brevibus, fasciis duabus antennisque (apice excepta) albidis; tibiis flavis.
Hab. Australia (Melbourne).

Very slender, elongate, deep glossy black; the prothorax moderately, the elytra strongly punctured, the latter very short, not exceeding half the length of the abdomen, the base and band at the apex a pale yellowish-white; antennæ white, except the three apical joints; tarsi yellow. Length 2 lines. British Museum.

LEMIDIA [Cleridæ].

Spinola, Rev. Zool. 1841, p. 75.

Lemidia carissima.

L. fulvo-testacea, nitida; elytris læte-viridibus, humeris, fascia media apiceque aurantiacis; tibiis tarsisque posticis nigris.
Hab. Australia (Melbourne).

Shining reddish-testaceous; elytra bright green, shoulders, band across the middle, and apex rich orange-red; eyes, tibiæ and tarsi black; throat, meso- and metathorax, and patch on the abdomen brassy-black. Length 3 lines.

Lemidia insolata.

L. pallide fulva, breviter setosa; prothorace nitido; elytris striato-punctatis, dense tomentosis; oculis apiceque mandibularum nigris.
Hab. Macassar.

Pale tawny, covered with short, erect, setose hairs; head and prothorax glossy; scutellum and elytra with a dense opake pale tomentum, the latter regularly and finely punctured; eyes and tips of the mandibles black. Length 5 lines.

TENERUS [Cleridæ].

Laporte de Castelnau, Silberm. Rev. Entom. iv. p. 43.

Tenerus telephoroides.

T. subangustatus, ater, nitidus ; prothorace, articulo basali antennarum, labro, tibiisque flavis.
Hab. Australia (Moreton Bay).

Rather narrow and depressed, black, shining, finely punctured, covered with short setulose hairs ; head scarcely as broad as the prothorax, black; oral organs and palpi yellow, except the tips of the mandibles, which are black ; prothorax reddish-yellow, the anterior border black, three mammillated prominences on the disc, placed transversely ; scutellum small, black ; elytra deep black, shoulders rather prominent ; femora and tarsi black, coxæ and tibiæ yellow; antennæ black, the basal joint yellow; body beneath black, shining, except the prothorax, which is yellow. Length 3 lines.

The joints of the antennæ are strongly produced on one side, as in the majority of the species of this genus, beginning from the third. I have only seen a single specimen, which is in my own collection.

CHORESINE [Cleridæ].

Head small, transversely triangular in front, slightly exserted behind. Eyes rounded, prominent, entire. Antennæ 11-jointed, linear, not half the length of the body, arising in front of the eyes; the first joint twice as long as the second, which is only a little shorter than the third, the fourth and fifth slightly longer, the rest subequal. Labrum transverse, entire. Mandibles strongly curved, bidentate at the apex. Palpi claviform, the joints very short and transverse, the maxillary much larger than the labial. Maxillæ rounded, two-lobed. Labium obovate. Prothorax subquadrate, constricted posteriorly before the base; pronotum distinct from the parapleuræ. Scutellum small, triangular. Elytra convex, nearly hemispherical, advancing at their insertion on the base of the prothorax. Legs slender; first joint of the anterior tarsi nearly covered by the second above; the middle and posterior tarsi with all the joints free, the three intermediate of all furnished with lamellæ. Abdomen slender, of six ? segments.

The habit of this very remarkable insect approaches in some respects the Melyrideous genus *Chalcas* ; the structure of the tarsi, however, is that of a Clerid, and although a very isolated form, I see no difficulty in placing it in the subfamily Enopliinæ.

Choresine advena. (Pl. II. fig. 2.)

C. flava ; elytris cyaneis; oculis pectoreque nigris.
Hab. Moluccas (Batchian).

Head and prothorax pure yellow ; scutellum and elytra dark indigo-

blue, covered with a sparse pale greyish pubescence; eyes and mesosternum black; rest of the body beneath, eyes and antennæ pale yellow.
Length 2 lines.

DOLIEMA [Tenebrionidæ].

Head short, transverse. Eyes lateral, contiguous to the prothorax, partially divided by the antennary orbit, larger below than above. Labrum
small, rounded in front. Mandibles thin, triangular, bidentate at the
apex. Antennæ short, perfoliate, moniliform, and gradually increasing
in thickness from the fourth joint to the seventh or eighth. Mentum
subquadrate. Labium small, entire; labial palpi stout, clavate, the
maxillary with its terminal joint subsecuriform. Maxillæ two-lobed, the
lobes ciliated (the inner armed * ?). Prothorax depressed, contracted
behind, broadly emarginate in front, its anterior angles rounded. Elytra
very depressed, parallel, abruptly bent down at the sides; the epipleural
plait narrow, terminating before reaching the apex. Legs short; coxæ
distant; tibiæ spurred, the anterior serrated externally; tarsi slender,
the first joint of the posterior as long as the last. Pro- and mesosterna
broad and flat, the former rounded posteriorly, and received into a slight
emargination of the mesosternum.

A remarkable genus, which might readily be taken for a *Platisus*, but which is very closely allied to, if not identical with,
Mr. Wollaston's *Adelina*. As, however, the characters of his genus
were drawn up from an insect which he suspects may not be congeneric with certain representatives in the British Museum of
M. Chevrolat's original, but unpublished *Adelina* (but which unquestionably includes the species now to be described), and his detailed
description differs in several, although somewhat secondary points,
from that given above, and he has taken no notice of the peculiar
elytra, I have thought it better to consider my species the type
of another group; and I do so with less hesitation, as the name of
Adelina has been long preoccupied by a genus of Gasteropods.
Doliema, thus restricted, has a remarkable range, *D. platisoides* occurring in Ceylon, Manilla, and the Moluccas, while a closely-allied
species, differing in nothing apparently but in having a somewhat
broader head, is found in the valley of the Amazons.

Doliema platisoides. (Pl. III. fig. 8.)

D. pallide ferruginea, nitida; capite modice transverso; prothorace postice bifoveolato.

Hab. Moluccas (Batchian); Ceylon; Manilla.

Extremely depressed, pale rusty testaceous, shining, and very mi-

* With a high power of the microscope, I cannot satisfactorily determine
whether the inner lobe of the maxillæ be armed or not.

nutely punctured; disc of the prothorax slightly concave, with two
large foveæ at the base; scutellum subquadrate; elytra punctured,
principally in rows of about six on each; eyes dark brown. Length
2⅓ lines.

ERYRPUS [Tenebrionidæ?].

Kirby, Trans. Linn. Soc. vol. xii. p. 389.

Eurypus cupripennis.

E. subangustus, subtilissime punctatus, cœruleo-chalybeatus, nitidus; elytris cupreis.
Hab. Brazil (Para).

Head rounded, pitchy, finely punctured; eyes and antennæ black;
prothorax narrower than the head or elytra, steel-blue, finely punctured, a deep transverse impression posteriorly; elytra elongate, gradually widening behind, rich coppery-red, minutely punctured; legs
small, pitchy; body beneath steel-blue. Length 5 lines.

Stilpnonotus eurypiformis (named, but not described, by Mr. G. R.
Gray in the English edition of the ' Règne Animal ') appears to me
to be referable to Mr. Kirby's *Eurypus*, a genus not alluded to by
M. Lacordaire in his great work. Mr. Kirby's species, *E. rubens*,
from the figure, seems to be a much broader insect than the present,
which it is not impossible may be identical with Olivier's *Tenebrio
nitens*. The pronotum is confounded with the parapleuræ, and the
anterior coxæ are contiguous and greatly exserted, two characters
which do not accord well with the Tenebrionidæ: the possession of
antennary orbits forbids its association with Lagriidæ. In habit it
is slightly assimilated to *Camaria*.

ŒDEMUTES [Helopidæ].

Head transversely subquadrate; epistome large, deeply inserted in front.
Labrum short, transverse, broadly emarginate. Eyes rather broad, sub-
lunate. Last joint of the labial palpi securiform, of the maxillary narrowly triangular. Antennæ very short, clavate, 11-jointed, the first
joint nearly concealed by the antennary orbit, the second short, third
longest, the rest gradually increasing in breadth to the seventh, which,
with the remainder, forms a sort of club. Prothorax transverse, slightly
convex, carinated at its sides, the base closely applied to the elytra.
Scutellum small. Elytra ovate, very convex. Legs rather short; anterior femora strongly toothed; tibiæ slightly curved; tarsi very short,
the last joint longer than the rest together. Prosternum received in a
notch of the mesosternum.

Very near *Sphærotus*, from which it differs in the antennæ and
legs, especially in the profemora toothed as in *Enoplopus*, and in

the form of the prosternum and its contiguity to the mesosternum. My specimen is the only one I have seen, and was obtained from a small collection sent to this country by Mr. Thwaites, the Superintendent of the Botanic Garden at Peradenia.

Œdemutes tumidus.   (Pl. II. fig. 4.)

Œ. æneus; capite prothoraceque modice punctatis; elytris elevatis, punctato-sulcatis.

Hab. Ceylon.

Brassy-brown; head and prothorax irregularly, but not closely punctured; elytra very gibbous, as if inflated, each with about seven rows of strongly sulcated punctures; body beneath paler and less brassy. Length 4 lines.

Camaria [Helopidæ].

Encycl. Méthodique, Ins. vol. x. p. 454.

Camaria spectabilis.

C. viridi-ænea, subiridescens; elytris punctato-striatis, interstitiis cupreo-vittatis, apice obtusis; tarsis chalybeatis; corpore infra viridi-aureo.

Hab. North China.

Brassy-green, somewhat iridescent; head and prothorax finely punctured, the former with a semicircular impression above the epistome (common apparently to the genus); scutellum small, rounded posteriorly, chalybeate blue; elytra very convex, punctate-striate, punctures minute, the interstices in certain lights showing a stripe of rich copper-red, the apex obtuse; femora and tibiæ finely punctured, varied with blue, purple, and gold; tarsi dark blue; labrum, palpi, antennæ, and eyes black; body beneath rich golden-green. Length 12 lines (♂), 14 lines (♀).

Elacatis [Melandryidæ].

Head broadly triangular, as wide as the prothorax. Eyes distinct, large, ovate, contiguous to the prothorax. Antennæ arising from beneath a narrow orbit, eleven-jointed, the two basal joints thick, the second shortest, the third to the eighth subequal, slender, the last three forming a short ovate club. Labrum small, rounded anteriorly. Mandibles short, with a single tooth in the middle. Palpi robust, claviform. Maxillæ with two ciliated lobes. Labium small, subcordate. Mentum transverse. Prothorax subquadrate, posterior angle emarginate, the parapleuræ distinct. Elytra as broad as the thorax, tapering posteriorly, the epipleural plait very narrow. Legs short; anterior coxæ conical, contiguous, their acetabula closed behind, the intermediate subapproximate, oblique, furnished with trochanters, the posterior transverse; tibiæ spurred; tarsi very slender, the first joint long, the penultimate entire; claws simple. Mesosternum narrow, truncate behind.

I have not placed this genus among the Melandryidæ without hesitation, on account of its antennary orbits, and its acetabula closed behind ; on the other hand, its parapleuræ, distinct from the pronotum, make its location in any other family still more difficult. Except the comparative shortness of the maxillary palpi, it agrees with the Melandryidæ in most of the characters given by M. Lacordaire, according also in form with some of its genera, without, however, being related to any of them. Like *Tetratoma*, it has the antennæ terminating in a club, but only composed of three joints. In the drawing the maxillary lobes are much too large, compared to their palpus.

Elacatis delusa. (Pl. II. fig. 5.)

E. griseo-testacea, punctulata ; elytris fasciis tribus dentatis, maculaque basali nigris.

Hab. Borneo (Sarawak) ; New Guinea (Dorey).

Greyish-testaceous, finely punctured, a short setulose hair arising from each puncture ; prothorax with three or four very minute teeth at the side, and a shallow transverse impression near the base ; scutellum long and narrow ; elytra with three black, toothed bands, the first often interrupted or replaced by a few spots ; a patch of the same colour, also sometimes broken up into spots, at the base near the scutellum ; antennæ and legs testaceous-yellow, more or less clouded with brown ; body beneath ferruginous, slightly tomentose. Length 1½-2 lines.

My New Guinea specimen agrees perfectly well with those from Borneo ; but they all vary a little in colour, some being darker than others, and the black band and scutellar patch being more or less interrupted. A second species, and a much finer one, from the Moluccas, is in the collection of W. W. Saunders, Esq.

Biophida [Melandryidæ ?].

Head moderately long, tumid in front, suddenly contracted behind into a narrow neck. Eyes distant, lateral, reniform. Antennæ arising close to the eye, filiform, half as long as the body, 11-jointed ; the second very short, the rest subequal. Labrum transverse, inserted below the line of the front. Labial palpi filiform ; the maxillary elongate, with the last joint narrowly securiform. Prothorax depressed, semicircular, as wide as the elytra behind, its parapleuræ distinct. Elytra depressed, rather broader behind. Legs moderate ; anterior and middle coxæ contiguous, the former conical and elongate ; tibiæ spurred ; tarsi slender, the first joint of the four posterior as long or longer than the rest together, the penultimate bilobed ; claws undivided, strongly toothed beneath.

This is another of those puzzling genera, of which there are so

many among the Heteromera; in its habit it resembles *Scraptia*; but
as the more important characters are those of Melandryidæ, and that
family is also one which contains several anomalous forms, it seems
less objectionable to place it in that group than in any other.

Biophida unicolor.   (Pl. III. fig. 4.)

B. fulvo-testacea, pube pallidiori vestita; prothorace bifoveolato; oculis
fere concoloribus.

Hab. Natal.

Entirely of a light-brownish testaceous colour, rather closely covered
with short stiffish paler hairs; a large fovea on each side of the prothorax near the posterior angle; scutellum transverse, rounded behind;
eyes a little darker.   Length 4 lines.

Ischalia [Pedilidæ?].

Head small, contracted behind, and narrowed anteriorly below the eyes.
Antennæ shorter than the body, linear, 11-jointed; second joint smallest,
the rest subequal. Eyes reniform. Epistome and labrum large, covering
the mandibles. Maxillary palpi robust, the last joint securiform; labial
much shorter, terminating in a broad triangular joint. Maxillæ short,
obtuse. Prothorax narrowed anteriorly, irregular above, its posterior
angles produced, the epipleuræ confounded with the pronotum. Elytra
broader than the prothorax, subparallel, bent at the side, and concave
on the disc, the epipleural plait narrow. Legs moderate, anterior
acetabula open behind; all the coxæ approximate, the anterior and
intermediate conical; tibiæ unarmed; tarsi short, first joint longer
than the rest together, the penultimate bilobed; claws simple.

I refer this genus doubtfully to Pedilidæ, notwithstanding that it
agrees in two characters which M. Lacordaire considers of high importance, viz. the anterior acetabula largely open behind, and the
complete contiguity of the posterior coxæ. The family, however, as
it stands at present, is not a satisfactory one, and its learned pro-
poser will probably see reasons for modifying it eventually.

Ischalia indigacea.   (Pl. III. fig. 6.)

I. cyaneo-violacea; antennis pedibusque nigris, illis articulis tribus ultimis
albis.

Hab. Borneo.

Deep violet-blue; head and prothorax very minutely punctured
(scarcely visible under a strong lens), the latter more or less irregular;
scutellum small, triangular; elytra coarsely punctured, rich violet-blue;
antennæ black, with the last three joints white; legs black; body beneath
black, with a slight bluish tinge on the breast. Length 3–4 lines.

The irregularity of the surface of the prothorax varies; in extreme cases it has the appearance of being shrivelled up by desiccation. The structure of the palpi and maxillæ will be seen in the figures; the labium and mentum unfortunately disappeared in dissection.

MACRATRIA [Pedilidæ].

Newman, Entom. Mag. vol. v. p. 377.

Macratria mustela. (Pl. II. fig. 7.)

M. fusca; tarsis (basi excepta), palpis antennisque fulvescentibus, his apicem versus infuscatis; scutello parvo.
Hab. Natal.

Dark brown, sparingly covered with a pale golden-yellow pile; head and prothorax finely punctured, the latter with the sides posteriorly nearly parallel; scutellum small, subtriangular; elytra very thickly punctured*, with a larger series of punctures arranged in closely set rows, which are divergent at the base; antennæ and palpi tawny, the former, except three or four of the basal joints, gradually becoming darker; legs dark brown; the tarsi, except the basal joint of the posterior, yellowish. Length 3 lines.

Macratria fulvipes.

M. nigra; pedibus (tibiis posticis exceptis), palpis antennisque fulvis, his apicem versus infuscatis; scutello magno.
Hab. Macassar.

Black, very sparingly covered with a pale golden-yellow pile; head and prothorax rather coarsely punctured, the latter with the sides gradually but very slightly contracting posteriorly; scutellum large, subquadrate; elytra finely punctured, a larger series in rows as in the last species; legs (except the posterior tibiæ), palpi, and antennæ tawny-yellow, the latter with the three or four terminal joints darker. Length 2¼ lines.

Macratria pallidicornis.

M. picea; antennis, palpis pedibusque (posticis exceptis) testaceis; capite fulvescenti.
Hab. Borneo.

Pitchy, very sparingly covered with a pale yellowish or greyish pile; head and prothorax slightly punctured, the latter somewhat ovate; scutellum indistinct; elytra punctured as in the last species, but with the pile more confined to the rows of punctures; antennæ, palpi, and four anterior legs pale testaceous; the posterior femora, except at the base, tibiæ at the base and apex, and basal joints of the tarsi, dark brown or nearly black; head tawny-yellow. Length 2¼ lines.

* It is rather the appearance of punctures caused by minute transverse wrinkles.

Macratria fumosa.

M. rufo-brunnea; pedibus (posticis exceptis), palpis antennisque fulvis, his apicem versus infuscatis; capite pedibusque posticis nigris.

Hab. India (Dacca).

Light reddish-brown, with a pale greyish pile; head and prothorax very finely punctured, the latter rather broad and somewhat ovate; scutellum subtransverse; elytra punctured, &c., as in the preceding; legs (except the posterior pair), palpi, and antennæ fulvous, the latter with the last three joints dark brown; head and hind legs black, except the extremity of the tarsi, which are pale yellow. The claws in this species appear to be broadly toothed at their base. Length 2¼ lines.

Macratria subguttata.

M. atra, nitida, sparse albo-hirta: elytris, singulo maculis duabus, fere obsoletis, albis.

Hab. Moluccas (Batchian).

Glossy black, with much-dispersed whitish hairs; each elytron with two rather indistinct white or somewhat ashy spots, one a little before the middle, the other the same distance beyond it; antennæ, palpi, and mouth pale yellow, the former gradually deepening towards the apex into black; tarsi pale yellowish, except the first joint of the posterior pair. Length 3 lines.

EMYDODES [Lagriidæ].

Head very small, rounded behind the eyes, then contracting into a neck, which is nearly immersed in the prothorax. Eyes large, oblong, emarginate, transverse, and approximating both above and beneath. Labial palpi very small; maxillary elongate, the terminal joint ovate, pointed. Antennæ robust, shorter than the body, arising close to the eye, the first joint tumid, the second very short, the third to the tenth thick, triangular, with a bifid prolongation at the apex of each on one side, the eleventh elongate-ovate. Prothorax slightly transverse, rounded anteriorly, twice the breadth of the head, but much narrower at the base than the elytra, its parapleuræ confounded with the pronotum. Elytra depressed, with a subovate outline, the epipleura strongly bent in beneath. Legs robust; anterior coxæ large, approximate, shortly cylindrical; tibiæ not spurred, the four posterior thickened in the middle; tarsi short, the penultimate joint subbilobed; claws undivided, slightly toothed at the base.

A very curious genus, which, if rightly referred to Lagriidæ (and of this I have little doubt), differs entirely in the remarkable structure of the antennæ, in which it somewhat resembles the Pyrochroidæ. From my solitary specimen, I cannot make sure that the anterior acetabula are closed: they appear to be so, however. As far as I can judge from the parts *in situ,* the mentum is subtriangular and the labium obcordate.

Emydodes collaris. (Pl. III. fig. 3.)

E. nigra, setoso-hirsuta; capite prothoraceque luteis.

Hab. Brazil (Para).

Black, covered with short stiff hairs; head dull reddish-yellow; prothorax thickly punctured, clear reddish-yellow; elytra coarsely punctured, each in ten rows; tibiæ with long stiff hairs. Length 3 lines.

IODEMA [Cantharidæ].

Head shortly triangular. Eyes round, prominent, entire. Labrum small, rounded anteriorly. Palpi slender; terminal joint of the labial ovate, of the maxillary subcylindrical. Antennæ short, linear, the joints slightly obconic. Prothorax transverse, narrowed in front. Elytra subdepressed, wider behind; the sides somewhat concave. Legs slender; tibiæ bicalcarate; penultimate joints of all the tarsi small, triangular.

Differs from *Cantharis,* with which only it is likely to be confounded, in the short penultimate joint of its tarsi: the claws appear to be undivided, from the close application of their two divisions.

Iodema Clarkii. (Pl. III. fig. 1.)

I. atra, nitida; elytris violaceis; tarsorum posticorum articulo primo albido.

Hab. Brazil (Organ Mountains).

Head and prothorax deep glossy black, sparingly punctured, especially the latter; scutellum narrowly triangular; elytra dark violet-blue, thickly and irregularly punctured; body beneath and eyes black; base of the first joint of the posterior tarsi whitish; spurs of the middle tibiæ, and all the claws, except at their apices, yellow. Length 4 lines.

I am indebted for my specimen to the Rev. Hamlet Clark, who took several individuals at Constancia, in the Organ Mountains.

ZONITIS [Cantharidæ].

Fabricius, Syst. Entom. p. 126.

Zonitis cyanipennis. (Pl. III. fig. 5.)

Z. angustus, glaber, ater; prothorace, scutello, femoribusque (apice excepta) luteis; elytris cyaneis, nitidis.

Hab. Australia (Melbourne).

Narrow, glabrous, shining; head black, very narrow, and produced anteriorly; prothorax reddish-yellow, much longer than broad; scutellum dull yellowish; elytra narrow, parallel, rather convex, dark indigo-blue; legs black, with the coxæ and femora (except at the apex) yellow; meso- and metasterna, abdomen, and antennæ black. Length 6 lines.

This has scarcely the habit of any European *Zonitis*, and still less of some depressed Australian species, of which the *Z. dichroa* of Germar may probably be taken as the type.

ECELONERUS [Anthribidæ].

Schönherr, Gen. et Spec. Curcul. tom. v. p. 163 (Supplem.).

Eceloncrus albopictus. (Pl. II. fig. 3.)

E. subcylindricus, fuscus nigroque varius, fascia dentata antica et punctis tribus discoideis prothoracis, lunulis duabus magnis maculariformibus, apiceque elytrorum albis.

Hab. Australia (Moreton Bay).

Subcylindrical, pitchy, with a short dark-brown tomentum mixed with black, and blotched with pure white; head shortly ovate, brown, slightly spotted with white; prothorax subrotund, very convex, thickly punctured, dark brown, with an irregular, toothed, white, band-like mark on its anterior margin; scutellum very small, white; elytra punctato-striate, the alternate interstices raised and spotted with black, a large white lunate patch, more or less spotted with brown, extending longitudinally on the middle of each elytron, with its convexity towards the suture, and extending externally to its margin, the apex also with a white patch of the same character; antennæ pitchy-brown, slightly hairy; legs brown, annulated with white; body beneath dull cinereous, the three middle abdominal segments having on each side an impressed hairy spot; mesosternum three-lobed posteriorly. Length 6 lines.

With this fine species of *Ecelonerus* I also obtained a specimen of *Dipieza Waterhousei*, Pasc., hitherto only known from Aru, unless indeed (as I have elsewhere suggested as regards the genus) the *Œdecerus** bipunctatus* of M. Montrouzier, from Woodlark Island, should be identical, in which case it will probably be found to be very generally distributed in those regions.

The subjoined is a list of the Australasian Anthribidæ, so far, I believe, as they have been described:—

Ecelonerus subfasciatus, *Hope.* Sydney, Melbourne, Moreton Bay.
—— insularis, *Hope.* Melbourne.
—— albopictus, *Pasc.* Moreton Bay.
Cratoparis callosus, *Schön.* (mihi invisus).

* There is a genus of Galerucinæ bearing this name (although incorrectly written *Œdicerus*) in Hügel's 'Reise durch Kaschmir,' 1842. p. 556.

Anthribus bispinus, *Erich.* Tasmania.
Basitropis peregrinus, *Pasc.* Port Essington.
—— ingratus, *Pasc.* Port Essington.
—— solitarius, *Pasc.* Moreton Bay.
Tropideres musivus, *Erich.* Tasmania.
—— albuginosus, *Erich.* Tasmania.
Aræcerus sambucinus, *MacLeay.*
Ethneca Bakewellii, *Pasc.* Melbourne.
Genethila retusa, *Pasc.* Moreton Bay.
Ancylotropis Waterhousei, *Jekel.* Moreton Bay.
Dipieza Waterhousei, *Pasc.* Moreton Bay.

Dysnos [Anthribidæ].

Pascoe, Ann. and Mag. Nat. Hist. ser. 3. vol. iv. p. 438.

Dysnos semiaureus.

D. breviter ovatus, fusco-tomentosus, obscuro aureo-varius; prothorace corpore non latiore; articulis duobus basalibus tarsorum nigris.
Hab. Moluccas (Batchian).

Shortly ovate or inclining to cylindrical, with an opake brownish-black tomentum, varied on the elytra with pale longitudinal patches of pale golden hairs; prothorax not wider than the elytra; first two joints of the antennæ and the legs ferruginous, the tarsi with the two basal joints black. Length 1½ line.

Smaller and proportionably shorter than *D. auricomus*, with the prothorax nowhere wider than the elytra. In my specimen, the subulate process terminating the last joint of the antennæ is absent, a character which may probably turn out to be sexual only.

Habrissus [Anthribidæ].

Pascoe, Ann. and Mag. Nat. Hist. ser. 3. vol. iv. p. 432.

Habrissus omadioides.

H. angusto-ovatus, fusco-tomentosus griseo-varius; tibiis tarsisque annulatis.
Hab. Singapore.

Narrowly ovate, with a tawny yellowish tomentum varied with dark brown; head tawny, with a longitudinal ridge between the eyes, and one on each side beneath them, not extending to the end of the rostrum; about five elongate indefinite marks on the prothorax; elytra striato-punctate, a large dark brown patch at the base and another in the middle common to both elytra, the alternate interstices also spotted with brown, particularly at the sides; legs very distinctly annulated with clear brown and tawny; body beneath greyish, inclining to ashy. Length 3 lines.

MISTHOSIMA [Anthribidæ].

Pascoe, Ann. and Mag. Nat. Hist. ser. 3. vol. iv. p. 434.

Misthosima lata.

M. late ovata, fusca grisco-varia ; pedibus brunneis, tibiis, apice, tarsisque (basi excepta) nigris.

Hab. Moluccas (Batchian).

Broadly ovate and very slightly depressed, pubescent, dark brown varied with a few spots of grey, principally on the elytra, the striæ have also a line of grey hairs in each ; antennæ about two-thirds the length of the body, ferruginous, the club nearly black ; legs pale brown, the tibiæ, at the apex, and tarsi, except at the base of the first joint, black. Length 2½ lines.

NESSIARA [Anthribidæ].

Pascoe (*Nessia*), in Annals and Mag. Nat. Hist. ser. 3. vol. iv. p. 329 ; non *Nessia*, J. E. Gray.

Nessiara planata.   (Pl. II. fig. 1.)

N. hirta, fusca, grisco-varia ; elytris deplanatis, retusis, singulo postice bituberculatis.

Hab. Moluccas (Batchian).

Clothed with short appressed dark brown hairs varied with grey, which are more or less ashy ; head entirely grey, the rostrum with a central carina, and a shorter one on each side below the eye ; prothorax with the sides dark brown spotted with grey, the disc with a central subtriangular ashy spot which is abruptly narrowed behind ; scutellum ashy ; elytra punctato-striate, rather broad, flatly depressed, suddenly bent down near the apex, the outer posterior angle of each bituberculate, the depressed portion dark ashy, the sides dark brown, the alternate interstices with black and pale yellowish-grey spots ; body beneath yellowish-brown ; legs annulated with dull brown and pale grey ; eyes pale brown, somewhat lustrous. Length 5 lines.

I have elsewhere mentioned my suspicions that this genus is synonymous with *Dendropemon*, Schön., and M. Jekel is inclined to take the same view of it ; as, however, the name was previously used by Perty, or what will be considered to amount to the same thing— for his orthography was *Dendropœmon*—another name must be adopted, and *Nessia* having been applied to a group of Saurians, I have thought a modification of it to *Nessiara* will be attended with the least inconvenience. *Stenocerus platipennis*! Montrou., is evidently nearly allied to the species just described, and his three other *Stenoceri* probably belong likewise to this genus. *S. Garnotii*, Guér., and the insect figured in the ' Voyage de la Bonite,' Coleop. pl. ii.

fig. 21, under the name of *Stenocere Damier*, are doubtless also *Nessiaræ*. *Nessiara centralis*, Pasc., is found in the Moluccas, as well as in Borneo.

BASITROPIS [Anthribidæ].

Jekel. Ins. Saundersiana, p. 90.

Basitropis solitarius.

B. elongato-subcylindricus, fusco-tomentosus; capite prothoraceque obscure griseo-variis; elytris striato-punctatis, interstitiis alternis elevatis, irregulariter albo-maculatis.
Hab. Moreton Bay.

Elongate, subcylindrical, with a short dark brown tomentum, slightly varied with greyish-white; head shortly ovate, eyes rather large; prothorax a little longer than wide, varied anteriorly and at the sides with greyish; scutellum minute; elytra punctate-striate, the alternate interstices raised and spotted with white, the spots a little before, as well as behind the middle, elongate, forming an indistinct, oblique, band-like mark; antennæ dark brown; legs paler, varied with greyish; body beneath greyish-brown. Length 3 lines.

This species, together with *B. peregrinus* and *B. ingratus* from Port Essington, described by me in a recent number of the 'Annals and Magazine of Natural History' (Dec. 1859, pp. 432, 433), &c., differ from *B. nitidicutis*, Jekel, the type of the genus, in their narrower and more elongate form, and their brown, not ashy, colour.

DINORHOPALA [Curculionidæ].

ead small, abruptly contracted below the eyes into a short rostrum. Eyes large, round, prominent. Antennæ short, straight, arising close to the eyes in a cavity formed between them and a short thick process, twelve-jointed, the first subpyriform, elongate, the second shorter, subcylindrical, the third to the eighth slender, gradually diminishing in length, the last four forming an ovate compact club. Prothorax subtriangular, lobed at the base, narrow anteriorly, irregular above. Elytra large, much wider than the prothorax at the base, very irregular and spinous. Anterior and intermediate legs moderate, the femora clavate and unidentate beneath, each tibia with a single curved spur; the posterior longer, their femora slender at the base, abruptly clavate at the apex, and armed with a strong tooth, their tibiæ strongly compressed and curved; the tarsi of all short, the penultimate joints broadly lobed; claws toothed beneath; anterior coxæ approximate, intermediate and posterior widely apart. Meso- and metasterna very large.

The affinity of this genus is no doubt with *Tachygonus*, and judging from its posterior legs, it is probably also saltatorial. As the import-

ance of the geniculation of the antennæ is now only recognized as a
secondary character, I think M. Jekel* has done good service in
referring all the groups of Schönherr's Orthocerati, after eliminating
those which evidently belonged to the true Curculionidæ, to four
families. *Tachygonus* is one of the genera so removed, and this
M. Jekel seems inclined to place near *Ceutorhynchus*.

Dinorhopala spinosa.   (Pl. III. fig. 2.)

D. atra, subnitida; rostro, antennis, pedibusque (clava tibiisque posticis
exceptis) fulvescentibus.

Hab. Burmah (Rangoon).

Glossy black; rostrum, throat, antennæ, the four anterior legs, bases
of the posterior femora and tarsi brownish-yellow. Length 2½ lines.

The figure, which is in no degree exaggerated, will give a better
idea of this singular little insect than the most lengthened description. It was taken, with other very interesting species, by an
English officer at the time of our recent occupation of Rangoon.

ORTHOSTOMA [Cerambycidæ].

Serville, Ann. de la Soc. Ent. de France, t. iii. p. 61.

Orthostoma cyanea.

O. læte-cærulea; thorace luteo; antennarum articulis tribus ultimis albis.

Hab. Brazil (Para).

Bright cobalt blue; head thickly punctured; eyes dark brown; prothorax reddish-yellow, finely punctured; scutellum subquadrate; elytra
minutely granulated, sparingly clothed with short stiff black hairs; a
few scattered hairs on the legs and antennæ; antennæ somewhat longer
than the body, the last three joints white; jugulum, prosternum, and
anterior coxæ yellow; abdomen glossy greenish-blue. Length 8 lines.

OSTEDES [Lamiidæ].

Pascoe, Trans. Ent. Soc. n. s. vol. v. p. 43.

Ostedes spinosula.

O. grisescens, fusco-variegata; prothorace trituberculato, lateribus muticis; elytris basin versus spinosis, spina incurva.

Hab. New Guinea (Dorey); Moluccas (Batchian).

Finely pubescent, greyish varied with brown; head small, deeply
sulcated in front; prothorax a little longer than wide, the sides unarmed, the disc with two broadly depressed tubercles towards the anterior margin; scutellum scarcely transverse, rounded behind; elytra
rather narrow, the basal half sparingly punctured, a prominent, strongly

* Insecta Saundersiana, pt. ii. pp. 156, 157.

recurved spine on each towards, but at some distance from the base, the sides with three or four brown patches, the outer apical angle produced; legs dark brown, the basal portions of the femora and tibiæ reddish-testaceous; antennæ longer than the body, slightly setose, reddish-brown, the apices of the intermediate joints black; body beneath reddish-brown. Length 5 lines.

From the slender and elongated tarsi, particularly the posterior, I should be inclined to refer this genus to the neighbourhood of *Œdopeza*, rather than to *Monohammus*, where formerly I had doubtfully placed it. Except the slightest possible variation in the patches on the elytra, there appears to be no difference between the Batchian and Dory insects.

Astathes [Lamiidæ].

Newman, The Entom. p. 299.

Astathes caloptera.

A. atra, nitida, breviter setosa; elytris læte cyaneo-violaceis; antennis testaceis, apicem versus infuscatis.
Hab. Borneo.

Ovate, sparingly clothed with short setose hairs; head and prothorax shining black with a slight copper tinge, and a few scattered punctures; scutellum very transverse, black; elytra deep bluish-violet, very bright and glossy, and in certain lights having a strong purple tinge, their disc somewhat concave, and each having two abbreviated costæ; antennæ pale testaceous-yellow, the apex dark brown; body beneath and legs black, the last abdominal segment obscurely testaceous. Length 5 lines.

A most beautiful species, approaching my *A. purpurea*, but perfectly distinct. It was found in Borneo by Lieut. De Crespigny; and does not occur, I believe, in Mr. Wallace's collections.

Euryptera [Lepturidæ].

(Encycl.) Serville, Ann. de la Soc. Ent. de France, t. iv. p. 222.

Euryptera albicollis.

E. nigra; prothorace, humeris, femoribusque subtus albis.
Hab. Brazil (Para).

Opake brownish-black, finely punctured; head narrowly elongate, the sides whitish, front between the eyes darker; epistome, labrum and palpi glossy black; prothorax white, with a yellowish tinge, a blackish spot on its anterior border; scutellum triangular, black; elytra nearly parallel, black, with a fine, scattered, greyish pubescence, which gives them a dull tinge, the shoulder with a triangular whitish spot, the apex truncate, its outer angle sharply spined: femora beneath, coxæ, and

base of the first joint of the intermediate tarsi whitish; antennæ with
the bases of all the joints, except the first two, white; breast and throat
white, rest of the body beneath smoky-black.   Length 8 lines.

TRIPLATOMA [Erotylidæ].

(Westw.) Lacordaire, Monog. des Erotyliens, p. 44.

Triplatoma Sheppardi.

T. elongato-ovata, subtilissime punctata, nigro-ænea; elytris singulis
maculis duabus luteis; pedibus ferrugineis, genubus tarsisque infuscatis.
Hab. Moluccas (Batchian).

Elongate-ovate, rather narrow, dark brassy-black, and very minutely
punctured above; elytra very convex, truncate at the apex, each with
a round yellow spot near the shoulder, and another towards, but at
some distance from, the apex (sometimes two similar spots on the prothorax anteriorly); legs glossy ferruginous, femora at the apex and
tarsi dark brown or nearly black; body beneath smooth, brownish,
with a slight brassy tinge.   Length 11 lines.

I have dedicated this fine and, I believe, hitherto undescribed
species to Edward Sheppard, Esq., F.L.S. &c., of Notting Hill, the
possessor of an extensive collection of Erotylidæ.

EXPLANATION OF THE PLATES.

PLATE II.

Fig. 1. *Nessiara planata.* Moluccas.
Fig. 2. *Choresine advena.* Moluccas.
Fig. 3. *Ecelonerus albopictus.* Moreton Bay.
Fig. 4. *Œdemutes tumidus.* Ceylon.
Fig. 5. *Elacatis delusa.* Borneo.
Fig. 6. *Sostea Westwoodii.* Borneo.
Fig. 7. *Macratria mustela.* Natal.
Fig. 8. *Cormodes Darwinii.* Lord Howe's Island.
Fig. 9. *Allelidea brevipennis.* Melbourne.

PLATE III.

Fig. 1. *Iodema Clarkii.* Organ Mountains.
Fig. 2. *Dinorhopala spinosa.* Burmah.
Fig. 3. *Emydodes collaris.* Para.
Fig. 4. *Biophida unicolor.* Natal.
Fig. 5. *Zonitis cyanipennis.* Melbourne.
Fig. 6. *Ischalia indigacea.* Borneo.
Fig. 7. *Byrsax cœnosus.* Singapore.
Fig. 8. *Doliema platisoides.* Moluccas.

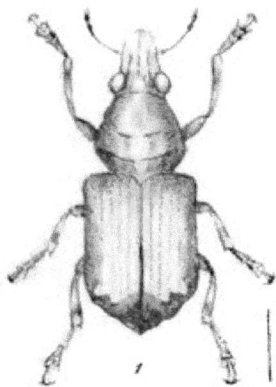

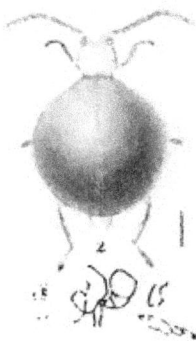

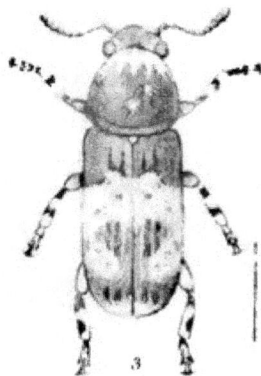

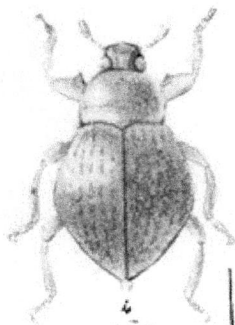

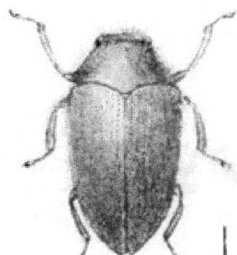

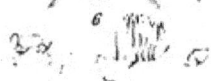

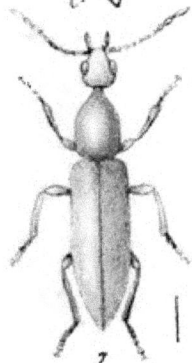

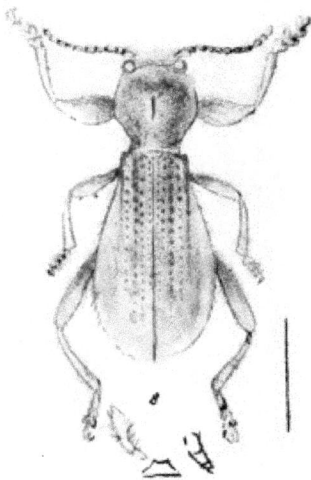

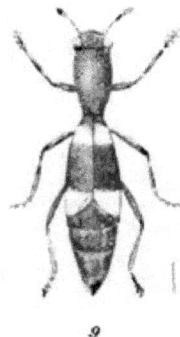

F.P.P. lith.

W. West imp

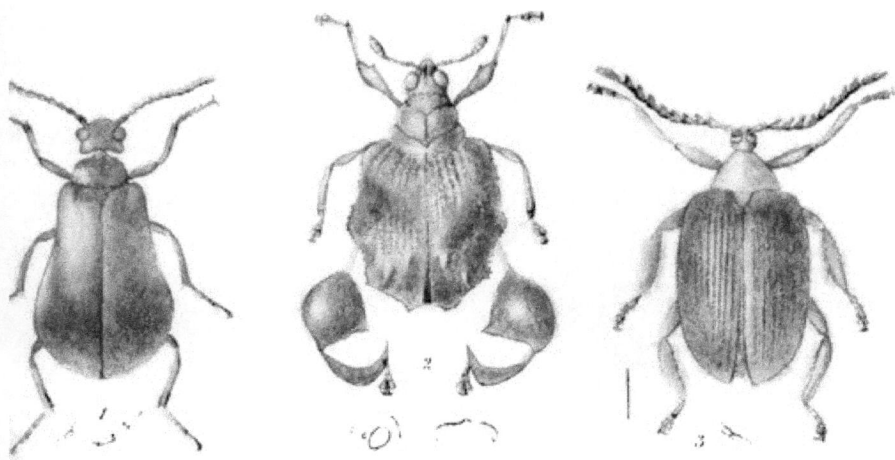

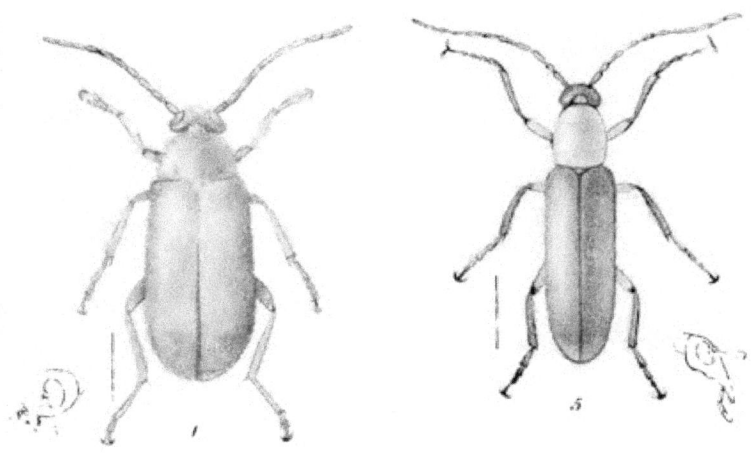

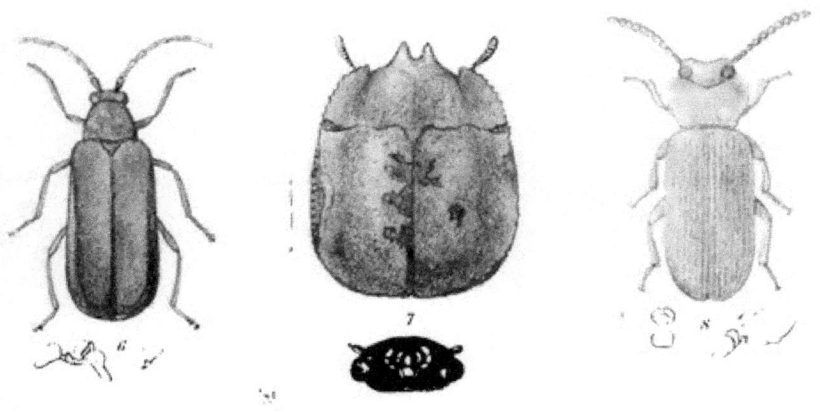

NOTICES

OF

NEW OR LITTLE-KNOWN

GENERA AND SPECIES OF COLEOPTERA.

BY

FRANCIS P. PASCOE, F.L.S., ETC.

[*From the* JOURNAL OF ENTOMOLOGY *for October.*]

Part II

Calonecrus [Nitidulidæ].

Thomson, Arch. Ent. i. p. 117.

Calonecrus rufipes.

C. rufo-flava; oculis elytrisque nigris.
Hab. Borneo.

Entirely reddish-yellow, except the eyes and elytra, which are black; head and prothorax finely, elytra more coarsely punctured; sides of the latter, pygidium, femora and tibiæ pubescent. Length 3 lines.

Proportionally a more slender form than *C. Wallacei*, Thoms., and altogether less robust, with the antennæ and legs reddish-yellow, and not black as in that species.

Prostomis [Cucujidæ].

Latreille, Fam. Nat. du Règne An. p. 397.

Prostomis morsitans.   (Pl. V. fig. 6.)

P. oblongus, testaceus vel piceo-testaceus; prothorace transverso; elytris punctato-striatis.
Hab. India (Darjeeling).

Larger and proportionally broader than *P. mandibularis*, the prothorax transverse, the antennæ shorter, &c. Length 4 lines.

In the only two specimens which I have seen (in the British Museum), one is very much darker than the other. Mr. Bakewell has another very distinct species from Melbourne.

Rhyssopera [Cucujidæ].

Head small, slightly exserted, narrowed anteriorly. Antennæ of moderate length, the first joint thick, abruptly contracted at its base, the rest more or less ovato-triangular, the last three stouter, forming a loose oblong club. Eyes transverse, rather prominent. Mandibles bidentate at the apex. Labrum long, narrow, rounded anteriorly. Palpi claviform, the last joint broadly ovate, obliquely truncate, the maxillary much larger than the labial, and widely separated at their origin. Mentum subquadrate, not larger than the labium, which is transverse and emarginate anteriorly; external maxillary lobe broad, strongly ciliated, inner very narrow. Prothorax subcordate, scarcely sinuated in front. Elytra much broader than the prothorax, parallel, slightly depressed. Legs small; anterior coxæ transverse, scarcely approximate; tibiæ biculcarate: tarsi five-jointed, slender, short, hairy beneath.

If rightly referred to the Cucujidæ, the position of this genus will be near *Silvanus*, which it approaches in habit and in its clavate antennæ.

Rhyssopera areolata. (Pl. VII. fig. 4.)

R. fusca, sparse flavo-pubescens; prothoracis basi latiuscula; elytris areolatis.
Hab. Tasmania.

Opake umber-brown, with a sparse yellowish or almost golden pubescence, especially on the head and prothorax, the latter about as broad as long, rounded at the side, produced into a short acute angle anteriorly and slightly contracted behind, with four tubercles on its disc; scutellum transverse; elytra with their external margins serrated, each with three rows of coarsely punctured hexagonal nearly equal cells, the walls of which are formed by narrow raised lines; labrum, palpi, and legs ferruginous. Length 4 lines.

Rhyssopera illota. (Pl. VII. fig. 4, trophi only.)

R. fusca, sparse griseo-pubescens; prothorace longiore, basi angustata; elytris subareolatis.
Hab. Australia (Melbourne).

Like the last, but the prothorax is longer and much narrower posteriorly, the lines bounding the areolæ and punctures less marked, and the pubescence of a greyer hue.

GLŒANIA [Trogositidæ].

Head small, rounded and dilated below the eyes, emarginate in front. The labrum entire. Antennæ short, eleven-jointed, the last three forming a subunilateral, compressed club. Eyes round, prominent. Mandibles entire at the apex, toothed in the middle. Palpi robust, with the terminal joint subcylindrical; maxillary lobes finely toothed, the inner narrow. Labium quadrate, slightly fringed. Mentum large, quadrate. Prothorax subquadrate, narrower anteriorly, broadly sulcated at the side, and slightly margined. Elytra scarcely broader than the prothorax, subdepressed, the sides nearly parallel. All the coxæ distant; femora broad, compressed; tibiæ dilated below, terminating in a series of small teeth; tarsi slender, slightly ciliated beneath, the basal joint minute, the second as long or longer than the third and fourth together; claws toothed at the base. Prosternum rounded behind; mesosternum depressed.

The Trogositidæ do not appear to have any very definite characters, if we except the minuteness of the first tarsal joint, and include genera varying very much in their form. Of the four subfamilies

into which M. Lacordaire divides them, the present genus must be
arranged in the same group with *Trogosita* proper.

Glœania ulomoides.   (Pl. VIII. fig. 9.)

G. fusco-picea, sublævigata; prothorace antice excavato; elytris seriatim
punctatis.

Hab. Brazil (Rio).

Rather depressed, dark pitchy-brown, nearly smooth and shining;
head and prothorax minutely punctured, the latter with a long V-shaped
excavation in front, with the side broadly and deeply grooved, the
groove bounded internally by a gradually elevated ridge, which anteriorly forms a well-marked angular process projecting slightly over the
head, the external border of the groove formed by a narrow uniform
line, parallel to, and very slightly removed from the margin of the prothorax; scutellum very transverse; elytra with about seven rows of
minute punctures on each, the shoulder with a short broad ridge gradually passing into the disc posteriorly; anterior and intermediate
tibiæ rounded and denticulate externally at the extremity, with the
posterior strongly spurred internally; body beneath scarcely punctured.   Length 3 lines.

LEPERINA [Trogositidæ].

Erichson in Germar, Zeitschr. für die Entom. v. p. 453.

Leperina adusta.

L. oblonga, picea, supra albido nigroque squamosa; elytris postice latioribus.

Hab. Australia (Melbourne).

Oblong, pitchy-brown, rather sparingly covered above with short,
round, whitish scales, varied with black; head and prothorax with
large, shallow, crowded punctures with a few white scales, which are
more closely arranged on the sides of the latter; scutellum triangular;
elytra becoming gradually broader behind for about two-thirds of their
length, with three elevated lines on each, a broad stripe of whitish
scales extending along the suture, giving off a transverse branch at the
base, another rather below the middle, and expanding again at the
apex; lip, palpi, antennæ, legs, and borders of the prothorax, and
elytra beneath ferruginous.   Length 4 lines.

Leperina cirrosa.

L. oblonga, picea, supra albo nigroque squamosa, fasciculisque elongatis
ornata; elytris parallelis.

Hab. Australia (Moreton Bay).

Oblong, pitchy-brown, covered above with white, and more or less
lengthened scales, occasionally collected into fascicles, and varied
with black; head and prothorax remotely and deeply punctured, with

small and mostly white scales, except on the sides of the latter, where they are drawn out into long, linear, curved laminæ, on each side a long fascicle of whitish hairs mixed with black, and nearly meeting on the median line anteriorly : scutellum triangular, with a tuft of erect white scales ; elytra parallel, the scales towards the suture principally white, but more or less black at the side, long and filiform at the base, and spatulate on the exterior margins, a fascicle of long, erect black scales on the middle of each near the suture, and posteriorly another of mixed black and white scales ; body beneath, legs, antennæ, and lip dark brown or nearly black. Length 4 lines.

In this curious species, the lines on the elytra are nearly covered by the longer and more densely set scales. In all the Australian and New Zealand *Leperinæ* which I have examined, I have never noticed any other than simple, undivided eyes.

Leperina lacera.

L. oblonga, picea, supra nigro-squamosa, albo varia, fasciculisque brevibus induta ; elytris lateribus rotundatis.
Hab. Australia (Melbourne).

Oblong, pitchy-brown, partially covered with short black scales, and sparingly varied with white ; head coarsely punctured, with two black fascicles between the eyes ; prothorax with a smooth elevated median line, the sides strongly and deeply punctured, above four short black fascicles anteriorly, the margins densely covered with long, white, appressed scales ; scutellum triangular ; elytra rounded at the sides, the scales almost entirely black, spatulate at the margins, with a single short black fascicle on each shoulder ; body beneath, legs, and antennæ dark ferruginous. Length 4½ lines.

Bitoma [Colydiidæ].
Herbst, Die Käfer, v. p. 25.

Bitoma serricollis.

B. depressa, fusca ; prothorace punctato utrinque bicostato, lateribus serrulatis ; pedibus rufo-ferrugineis.
Hab. Australia (Melbourne).

Depressed, dark brown ; head coarsely punctured, grooved at the side below the eyes, and somewhat three-lobed anteriorly ; prothorax transversely subquadrate, coarsely punctured, with two costæ on each side, the exterior crenate, continuous with its fellow in front, the sides strongly serrulate, the anterior angle produced ; elytra a little wider than the prothorax, with five narrow costæ on each, the intervals transversely plicate from a double row of deeply impressed punctures ; antennæ and legs rusty-red ; body beneath coarsely punctured. Length 2 lines.

A little broader and more depressed than *Bitoma crenata* ; but, as far as external characters go, there can be no doubt as to its genus.

Bitoma proluta.

B. lata, depressa, fusca luteo varia; prothorace transverso, granulato, utrinque bicostato, costa interiori postice duplicata, antice emarginato, lateribus crenulatis.

Hab. Moluccas (Batchian).

Broad and depressed, dark brown varied with reddish-yellow; head punctured, a little concave on each side below the eyes; prothorax transverse, finely granulated, broadest at the base, rounded and dilated at the sides and irregularly crenate, deeply emarginate in front, the disc with two costæ on each side, the interior approximating and forming a short canal open towards the head and a loop posteriorly; elytra not wider than the prothorax, with five crenulated costæ on each, the intervals with a double row of deeply impressed punctures, a yellowish spot on the shoulder, another near the apex, between these three others, which, with their fellows, form an indistinct ring; legs pale yellowish-brown; body beneath dark brown. Length 2½ lines.

A broader species than the last, with the prothorax especially dilated at the sides and deeply emarginate anteriorly; hereafter it may be found necessary to separate it generically from *Bitoma*.

Bitoma jejuna.

B. angusta, rufo-brunnea; prothorace quadrato, granulato, utrinque tricostato, costa interna antica abbreviata.

Hab. Brazil (Rio).

Narrow, slightly depressed, reddish-brown, the elytra paler; head granulated, principally between the eyes; prothorax quadrate, equal in length and breadth, with three costæ on each side, the inner very short and confined to the anterior part, the interstices strongly granulated, the margins crenulated; scutellum subquadrate; elytra with five costæ on each, the interstices with two rows of rather shallow punctures; legs and antennæ ferruginous; body beneath dark brown, the abdomen reddish-pitchy. Length 1½ line.

Collected by Alexander Fry, Esq., to whose kindness I owe my specimens.

COLOBICUS [Colydiidæ].

Latreille, Gen. Crust. et Ins. ii. p. 9.

Colobicus parilis.

C. oblongus, nigro-piceus, sparse albido-setulosus; elytris punctato-striatis; antennis pedibusque ferrugineis.

Hab. Moluccas (Batchian).

In size and outline very like C. *emarginatus*, but the head is narrower and the form rather more convex; the colour on the head, prothorax, and elytra is uniform, with a pitchy gloss, not nearly opake, and the punctures are decidedly smaller, with the rows more approximate. Length 2 lines.

RECHODES [Colydiidæ].

Erichson, Naturg. der Ins. Deutschl. iii. p. 255.

Rechodes verrucosus.

R. modice convexus, fuscus; elytris antice subgibbosis, tuberculis oblongis disco instructis.

Hab. Natal.

Moderately convex, dark brown, more or less clouded with a lighter shade, or even inclining to grey; head with a line of four tubercles between the eyes, the antennary orbit large, a semicircular impression above the epistome; mentum large, quadrate; labium transverse, entire, ciliated in front; prothorax very transverse, wider than the elytra, the sides strongly dilated and margined with a double series of equal serriform tubercles, and deeply sinuated in front for the reception of the head, the disc with a row of five tubercles on each side the central line, the anterior pair accompanied by two others placed on the edge of the prothorax; scutellum small, quadrate; elytra seriato-punctate, slightly gibbous at the base, so as to be above the line of the prothorax, a row of small tubercles along the side, above this another of three oblong tubercles, followed by a third row which is incomplete in the middle, and lastly close to the suture is a line of smaller tubercles running, with a slight interruption posteriorly, to the apex,—the sides less strongly dilated than in the prothorax, but edged with a double row of serriform tubercles of the same size (in some specimens there is a lighter shade posteriorly, forming a band-like mark); antennæ, palpi, and eyes ferruginous, with a paler pubescence; body beneath dark brown, covered with small tubercles. Length 3 lines.

Rechodes fallax.

R. fere convexus, fuscescens; elytris antice subdepressis, tuberculis oblongis instructis.

Hab. Natal.

Closely allied to the former, but is smaller, less convex, the elytra narrower, and their base being depressed, they are on the same line with the prothorax; the disposition of the tubercles is almost precisely the same, except perhaps that they may be a trifle less marked; the colour in both species is somewhat variable. Length 2½ lines.

Rechodes signatus.

R. subdepressus, fuscus; prothoracis lateribus, elytrorumque macula magna albescentibus.

Hab. Natal.

Rather depressed, dark brown, tomentose; sides of the prothorax, and a large patch on the disc of the elytra, which, commencing at the base, is contracted in the middle and again expanded behind, and a smaller spot at the apex, greyish-white; disposition of the tubercles (which are all more or less conical) nearly as in the last; antennæ, palpi, and legs dull reddish-brown; under surface dark brown, covered with numerous small tubercles, and but slightly pubescent. Length 2¼ lines.

The few characters which Erichson has given of *Rechodes* accord perfectly well with the insects described above, except that the last joint of the maxillary palpi is scarcely securiform, although very broad and truncate. *Rechodes* is closely allied to *Ulonotus* and *Endophlœus*. To the former of these genera, M. Lacordaire refers, and I think correctly, *Bolitophagus antarcticus*, White; and I would also refer to it *Asida serricollis*, Hope. The genus *Pristoderus* of the latter author, founded on the *Dermestes scaber*, Fab., is probably identical with *Ulonotus*.

DISTAPHYLA [Colydiidæ].

Head small, transverse, scarcely visible from above, slightly dilated below the eyes, with a broad antennary groove beneath. Antennæ short, stout, 11-jointed, the two basal incrassated, the third longer than the rest, which are very transverse, the last two forming a short compressed club. Eyes large, round. Mandibles bidentate at the apex. Palpi robust, the terminal joint of the maxillary elongate, subcylindric, of the labial obovate: maxillary lobes narrow, ciliated. Labium very small, subcordate, fringed with long cilia. Mentum large, narrowed in front, rounded and dilated at the sides. Prothorax nearly quadrate, very irregular anteriorly, the margin granulate and setose. Elytra elongate, subcylindrical. Legs short; coxæ not contiguous; tibiæ gradually enlarging at the extremity, terminated by two small spurs, and bordered externally with a row of stiff setæ; tarsi with the three basal joints short, hairy below. Prosternum rounded posteriorly, the mesosternum depressed.

Judging from the position which Erichson has assigned to his genus *Phlœonemus*, this must be a near ally, although it cannot be by any means likened to *Colobicus*.

Distaphyla mammillaris.   (Pl. VIII. fig. 4.)

D. subcylindrica, picea (vel rufo-brunnea), fortiter punctata, setosa; prothorace antice bigibboso.
Hab. Brazil (Rio; Para).

Subcylindrical, pitchy-brown (or, in the Rio specimens, reddish-

brown), strongly and deeply punctured, the intervals having the
appearance of granulations, and being furnished here and there with
short stiff yellowish hairs or setæ; head deeply and semicircularly
grooved between the eyes; prothorax narrowing slightly behind, the
sides strongly granulated in a double row which is divided from the
granulations of the disc by a smooth line, anteriorly two large oblong
lobes overhanging the head, separated from each other by a narrow
groove, but posteriorly from the rest of the prothorax by a broad deep
hollow, which extends beneath them; scutellum small, triangular;
elytra with about eleven rows of large deep punctures; legs reddish-
ferruginous, with stiff scattered hairs; antennæ short, not longer than
the breadth of the head, dark brown, slightly setose; body beneath
roughly punctured. Length 2¼ lines.

Acropis [Colydiidæ].

Burmeister, Gen. Ins. no. 25.

Acropis Fryi.

A. rufo-picea, fulvescenti-hirta; elytris subseriatim tuberculatis, tuberculis
setiferis, fasciculis sextis nigris in medio obsitis; pedibus ferrugineis
nigro variis.
Hab. Brazil (Rio).

Reddish-pitchy, rather sparingly clothed with short, scale-like, greyish-yellow or almost golden hairs; head and prothorax with a few greyish setæ, the latter with about five dark spots on its disc; scutellum
rounded behind, closely covered with white hairs; elytra uneven,
with several small granular tubercles, ranged in more or less interrupted lines, each tubercle bearing at its apex a black erect rigid seta,
in the centre six dense fascicles of stiff black hairs, the first and third
of these nearer the suture than the second, an oblique stripe (composed
of more closely set hairs) below each shoulder, and towards the apex
another oblique patch of pure white hairs (composed, however, of two
distinct spots); legs dark ferruginous, with scattered grey hairs, the
femora varied with black, the tibiæ with a black ring in the middle;
antennæ and palpi pitchy-ferruginous; body beneath pitchy-brown with
pale greyish hairs. Length 3 lines.

This appears to differ from *A. tuberculifera*, Burm. (which, however,
I have not seen) in its larger size, the black fascicles, the yellow,
almost golden, tinge of its scale-like hairs, the absence of the shining
chestnut colour of the apices of the tibiæ, knees, tarsi, &c. Burmeister in his description of this genus has overlooked the basal joint
of the antennæ, and describes the second (last) joint of the club as
composed really of two, soldered together, and in this he is followed
by M. Lacordaire. I can find no trace of any such union, which, if
it existed, would give twelve joints to the antennæ, and not eleven,

as is really the case, that is to say, with the addition of the basal one.   *A. Fryi* and *A. incensa* were both taken by Mr. Fry at Rio.

Acropis incensa.

A. rufo-picea, fulvescenti-hirta; elytris subseriatim tuberculatis, tuber-culis setiferis, fasciculis plurimis fuscis in medio obsitis; pedibus fer-rugineis.

Hab. Brazil (Rio).

Differs from the last in its much smaller size, comparatively narrower and longer elytra, in the more numerous tubercles, and brown fascicles of hairs, the almost unvarying hue of the pubescence, although near the shoulder and apex may be traced rather more densely set patches of ·hairs than elsewhere, and the more uniform colour of the legs. Length 1⅔ line.

Acropis aspera.   (Pl. VI. fig. 1.)

A. nigra; prothorace granulato; elytris seriatim tuberculatis, setiferis, macula alba pone humeros, postice fasciculo nigro indutis; tibiis tarsis-que ferrugineis.

Hab. Brazil (Para).

Black, very slightly shining, and nearly free from pubescence, except two small patches on the anterior margin of the prothorax, and a short oblique white stripe, which, however, may be resolved into three spots, below the shoulder: scutellum rounded behind, naked; prothorax covered with small flat granulations; elytra with a large fascicle of black hairs on the lower third of each, the tubercles varying in size, but all furnished with a rigid black seta; antennæ, tibiæ, and tarsi ferru-ginous.   Length 2 lines.

LEMMIS [Colydiidæ].

Head vertical, rounded in front, and prolonged at the sides into two short peduncles bearing the eyes.   Antennæ short, eleven-jointed, the last two forming a short ovate club.   Prothorax short, very transverse, narrower behind, broader than the head anteriorly, the sides strongly denticulate.   Elytra nearly regular above, not broader, except at the base, than the prothorax.   Legs slender, first tarsal joint scarcely longer than the second.

The other characters of this genus are the same as those of *Acropis*, to which, indeed, it is nearly allied; the form, however, of the pro-thorax, added to the apparent absence of asperities, and the peculiar scaly crust, which covers the whole of the upper surface, as if a layer of opake varnish had been applied to it, obviously prevent its union with that genus.   The shortness of the first tarsal joint, being more of a comparative character, is, perhaps, of less importance.

Lemmis cœlatus. (Pl. VIII. fig. 3.)

L. oblongus, grisescens, setis hamatis brevissimis obsitus ; antennis capite brevioribus.

Hab. Brazil (Rio).

Oblong, brown ?, covered above as well as beneath with a scaly crust of a pale yellowish or greenish grey, with very short hooked hairs, particularly on the margins of the prothorax and elytra, curving forwards on the former, and backwards on the latter ; head (including the peduncles) narrower than the prothorax, this with seven well-marked but obtuse teeth on each side ; scutellum punctiform ; elytra a little wider posteriorly, each with three very slightly raised gibbosities near the suture, another at the shoulder, and externally towards the apex two or three more, but which are considerably less prominent ; antennæ pitchy, shorter than half the length of the head ; legs pitchy ; eyes dark brown. Length 1½ line.

In one of the two specimens now before me, the hairs are scarcely evident even on the margins, being, apparently, more enveloped by the scaly layer described above. In Mr. Fry's collection.

ETHELEMA [Colydiidæ].

Head vertical, rounded anteriorly, and prolonged at the side into a short peduncle bearing the eye. Antennæ as in *Acropis*, but more robust. Labium short, transverse, fringed with long hairs. Maxillary palpi robust, the terminal joint short, stout, obliquely truncate ; the labial with the two basal joints small, the third large, broadly subovate, slightly truncate. Mentum quadrate, very large. Prothorax as broad as the head, transverse, regular and convex above, narrowed anteriorly, the sides margined. Elytra oblong, nearly parallel, the surface smooth and regular. Legs rather slender ; tibiæ not ciliated externally, terminated by two short spines. Prosternum produced behind.

The above include the characters which, combined with the total absence of tubercles, chiefly separate this genus from *Acropis*.

Ethelema luctuosa. (Pl. VIII. fig. 6.)

E. oblonga, hirta, nigra, flavescenti-varia : prothoracis marginibus denticulatis, setosis.

Hab. Brazil (Rio ; Para).

Oblong, closely covered above with short scale-like black hairs, many of which are curved backwards, more or less varied with pale yellowish or white ; head not wider than the prothorax, a transverse depression in front below the peduncles ; prothorax scarcely narrower than the elytra, except at the base, the margins denticulate, each denticulation with a short curved hair arising from its apex ; scutellum very transverse ; elytra regular, punctate-striate, the striæ rather remote, the patches of yellowish hairs more conspicuous on the head and prothorax, but indefinite as to outline and varying apparently in different individuals; body beneath black; legs with a few scattered hairs only. Length 2 lines.

DASTARCUS [Colydiidæ].

Walker, Ann. and Mag. Nat. Hist. 3 ser. ii. p. 209.

Dastarcus confinis.   (Pl. VI. fig. 6.)

D. elongato-ovatus, fuscus; prothorace elytrisque costatis, costis ferrugineo-hirtis.

Hab. New Guinea (Dorey).

Elongate-ovate, dark brown, with stout, stiff, dilated, pale rusty hairs (or scales), which are chiefly confined to costæ and other elevations on the upper surface; head small, partially retracted in repose; prothorax with two waved grooves on each side, the outer smallest, and fringed with stiff hairs; scutellum scarcely visible; elytra punctato-sulcate, the costæ between them closely covered with stiff hairs; body beneath coarsely punctured, with a setaceous hair in the centre of each; palpi ferruginous. Length 5 lines.

Larger and stouter in proportion in all its parts than the Ceylonese *D. porosus*, but otherwise very closely allied.

I am unable, at present, to give any oral details of this curious genus, which Mr. Walker has only very briefly characterized, at the same time associating it with the Hydrophilidæ; it is, however, an undoubted Colydian, and evidently nearly allied to *Emmaglœus* of M. Léon Fairmaire. The large primo-abdominal segment and distant posterior coxæ suggest also an affinity with *Bothrideres* and *Derataphrus*; but its head, vestiture, and habit altogether, point to a distinct subfamily. It may be mentioned that all the coxæ are widely apart; the femora canaliculate beneath for the reception of the tibiæ, which are fringed with stiff hairs externally, and the anterior terminated by two spines, the inner of which is much longer and curved, whilst the outer, under a strong lens, is seen to be tridentate; the mouth is almost entirely closed below by the prolonged mentum? (as in *Derataphrus*), the small, pointed maxillary palpi protruding at the sides.

BOTHRIDERES [Colydiidæ].

Bothrideres succineus.   (Pl. V. fig. 3.)

B. niger; prothoracis angulis anticis subacutis, ecostatis; elytris striatis, tuberculatis, medio succineo-granulatis.

Hab. Brazil (Rio; Para).

Dull black, opake; head covered with rather distant, shallow punctures; prothorax remotely punctured, longer than broad, considerably narrower behind, its anterior angles not produced although somewhat acute, a tubercle at the side, the disc very concave anteriorly, with a deeply impressed, interrupted ring in the centre, behind which is an oval depression terminating posteriorly in an elevated tubercle, which again has on each side a short but very deep and narrow groove; elytra elongato-ovate, broader than the prothorax, deeply and irregularly striated, the interstices, except the two sutural on each side, with very strong, elevated, compressed tubercles, particularly at the base and inner row, becoming smaller and more conical externally,—each elytron, before the middle and on the outside of the second sutural stria, with two pellucid granules of an amber colour; body beneath with rather shallow, large, and somewhat remote punctures. Length 2½ lines.

The upper part of the labium in the figure is intended to represent its cilia: as it stands, it only shows their position.

Bothrideres latus.

B. niger, latior; prothoracis angulis anticis productis, utrinque tricostatis. *Hab.* Brazil (Santarem).

Wider than the last, black, opake; head rather coarsely and deeply punctured; prothorax less coarsely punctured, rather wider than long, emarginate in front to receive the head, its anterior angles slightly produced, with three strong ribs on each side, the inner occupying the anterior half only, the outer terminating in the anterior angle, the disc largely impressed with a bilobed protuberance in the centre, and opening out behind into a deep channel, which is bounded on each side by an oblique protuberance; elytra broader than the prothorax, strongly ribbed, the interstices with shallow, somewhat remote punctures, the ribs seven on each elytron, the external and the two sutural ones less marked than the others; antennæ not longer than the breadth of the head; palpi ferruginous; body beneath remotely punctured. Length 3 lines. British Museum.

Sosylus [Colydiidæ].

Erichson, Natur. der Insekt. Deutschl. iii. p. 288.

Sosylus sulcatus. (Pl. VI. fig. 1.)

S. niger, subnitidus; prothorace medio lineolato; elytris apice obtusis, in singulo quadrisulcatis. *Hab.* Brazil (Para).

Black and slightly shining; head finely punctured, regular, a little convex in front; prothorax oblongo-ovate, twice as long as the head, finely punctured, a very delicately elevated line along the middle, terminating posteriorly between two short linear impressions; scutellum very narrow; elytra nearly parallel, obtuse at the apex, each with five elevated costæ having between them four broad deep grooves, the two outermost costæ uniting posteriorly and forming a slightly projecting angle at the apex ; antennæ and legs dark ferruginous, shining ; body beneath shining, dark reddish-brown, with small oblong impressed spots. Length 4 lines.

ANARMOSTES [Colydiidæ].

Head subquadrate. Antennæ short, eleven-jointed, the two basal incrassated, the third longest, the rest gradually decreasing in length to the eighth, the last three forming an ovate, compressed, perfoliate club. Eyes large, round, slightly divided in front. Maxillary palpi subcylindric, the last joint obliquely truncate, the labial smaller, subacuminate. Prothorax elongate, narrower posteriorly, deeply sulcate, not contiguous to the elytra. Scutellum punctiform. Elytra elongate, nearly parallel, ribbed, wider than the prothorax. Legs short ; coxæ not contiguous ; tibiæ spurred, somewhat dilated and more or less toothed externally near the apex ; tarsi slender, hairy beneath, the basal joint subelongate. Prosternum prominent, keeled in the middle. Abdominal segments gradually diminishing in size.

Allied to *Sosylus*, with which it also agrees in habit, but at once distinguished by its triarticulate club and sulcate prothorax. I have not dissected the mouth of my specimen (which I owe to the kindness of Mr. Fry, by whom alone, I believe, it has been taken) ; but the mentum seems to be very small, and attached internally to the large subquadrate jugular plate, which M. Lacordaire has, apparently, denominated the *"sous-menton"* ; the point of insertion of the palpi is, however, not covered by it, but is more than usually obvious.

Anarmostes sculptilis.   (Pl. VIII. fig. 8.)

A. elongatus, piceo-fuscus : pedibus rufo-piceis.
Hab. Brazil (Rio).

Elongate, dark pitchy-brown ; head and prothorax covered with numerous impressed punctures, with a very short hair-like point in the centre of each, the latter with five deep longitudinal grooves ; scutellum hollowed out in the middle ; elytra about three times the length of the prothorax, each with five strongly marked costæ, the intervals with a double row of elongated punctures, giving the spaces between them a granulated appearance : antennæ much shorter than the prothorax, yellowish-red ; legs dark pitchy-red ; tibiæ finely ciliated and armed externally at the base with three or four teeth : body beneath coarsely punctured, the abdominal segments with numerous fine, longitudinal, but more or less interrupted lines. Length 4½ lines.

ASPROTERA [Colydiidæ].

Head rather narrow, depressed, slightly expanded at the sides over the antennæ. Eyes large, round, with a deep antennary groove beneath. Antennæ short, ten-jointed, the first two incrassated, the remainder to the ninth more or less transverse, the tenth forming a round compressed club. Labrum small, entire. Palpi rather short, filiform, the last joint subcylindric. Mentum very transverse. Prothorax elongate, with nearly parallel, slightly margined sides, constricted a little at the base, produced anteriorly into a broad lobe overhanging the head. Elytra lengthened, parallel, very convex. Legs short; posterior coxæ distant: femora strongly grooved beneath for the reception of the tibiæ: tibiæ enlarged at their extremity, without spurs, ciliated on their external margin: tarsi slender, the three basal joints very short. Prosternum produced. The first two abdominal segments larger than the others.

Although the second abdominal segment is fully as large as the first, yet, as they exceed the remainder, this genus cannot be placed in any group in which the segments are equal; otherwise, as its posterior coxæ are not contiguous, it might be associated with *Pycnomerus, Apeistus,* &c. In its scaly pubescence it differs from *Bothrideres, Sosylus,* and all the genera of that group (and the character, as well as the absence of vestiture, like the sculpture, appear to me to be of importance in this family). The antennæ, described as ten-jointed, may probably have eleven, the club being composed of two, soldered together. In the figure eleven joints are given, but the third should be united with the second.

Asprotera inculta. (Pl. VI. fig. 3.)

A. elongata, cylindrica, fusca, supra albido-squamulosa; elytris seriatim punctatis, interstitiis squamulosis.
Hab. Natal.

Elongate, cylindrical, dull brown, furnished above with stiff whitish scale-like setæ; head coarsely punctured, with few scales; prothorax strongly and thickly punctured, with numerous scales between them, the anterior margin on each side obliquely grooved; scutellum very small; elytra very coarsely seriato-punctate, the alternate interstices with a more closely set row of scales than the intermediate ones; antennæ not longer than the breadth of the head, reddish-brown; legs reddish-brown; body beneath dark brown, coarsely punctured. Length 3½ lines.

PENTHELISPA [Colydiidæ].

Head small, slightly dilated below the eyes. Antennæ short, stout, eleven-jointed, the last two forming a short ovate club. Eyes round. Mandibles bidentate at the apex. Maxillary palpi robust, the terminal joint broadly ovate, the labial smaller. Maxillary lobes short, ciliated, somewhat falcate, the inner narrower. Labium very transverse, rounded anteriorly, and finely ciliated. Mentum subquadrate, its anterior angles rounded. Prothorax subquadrate, scarcely emarginate in front, with a narrow margin at the side. Elytra elongate, subparallel. Legs short; coxæ distant; tibiæ smooth externally, dilated at the extremity, and terminated by two or three spurs; tarsi stout, the first three joints subequal. Abdominal segments equal. Prosternum continuous with the mesosternum.

I believe this genus will be found to include that portion of Erichson's *Pycnomerus* which is characterized by its eleven-jointed antennæ. *Dechomus*, distinguished by having eight only, has been recently separated by M. Jacquelin du Val. The two European species, *P. terebrans* and *P. inexspectus*, with ten joints, will, therefore, alone represent the true *Pycnomeri*. The species described below has very slightly impressed antennary grooves, a character which, among the Pycnomerinæ, does not appear to be of generic importance.

Penthelispa porosa.

P. elongata, subdepressa, rufo-picea; prothorace fortiter punctato; elytris punctato-striatis.
Hab. Brazil (Rio).

Elongate, subdepressed, reddish-pitchy; head slightly convex in front, moderately punctured; prothorax longer than broad, a little narrowed posteriorly, covered with large and somewhat remote punctures; scutellum indistinct; elytra coarsely striato-punctate, the striæ very narrow, with the punctures oblong; legs smooth, the internal border of the tibiæ towards the extremity, especially of the anterior, slightly spinulose; body beneath pitchy-brown, with large shallow punctures. Length 2 lines.

Hyberis [Colydiidæ].

Head short, transverse, immersed in the prothorax nearly to the eyes. Antennæ of moderate length, arising beneath the lateral border of the head, moderately thick, ten-jointed, the joints ovate-elongate, setigerous, the first rather incrassated, the third longest, the tenth forming a pyriform club. Eyes lateral, round, rather prominent. Mentum nearly quadrate. Palpi claviform, terminal joint of the maxillary much larger than the others, shortly ovate, truncate, of the labial oblong-ovate. Prothorax transverse, bisinuated in front, rounded and strongly serrated at the side, narrowed behind. Elytra much wider than the prothorax, broadly ovate, convex. Legs moderate; coxæ distant; femora robust; tibiæ fusiform; tarsi short, the basal joint longer than the two following. Abdominal segments nearly equal.

As the only specimen I have seen of this insect belongs to the British Museum, I am unable to give any account of its oral organs; but there can be no doubt that it is nearly allied to *Apeistus*, and it would therefore be interesting to know if it be furnished with paraglossæ, as in that genus. It is remarkable that the basal joint, which in *Apcistus* is very indistinct, and was considered to be a mere knob (and the insect, therefore, trimerous) by Erichson, should be also in *Hyberis* so indented, that when viewed sideways it seems composed (at least in the intermediate tarsus) of two distinct joints; but the absence of any division beneath shows that it is not really so.

Hyberis araneiformis. (Pl. VII. fig. 1.)

H. fuscus, tuberculiferus, fulvo-setosus; antennis capite prothoraceque longioribus.

Hab. Borneo.

Broadly ovate, dark brown, opake, covered with small tubercles and short stiff fulvous hairs: head scarcely more than half the breadth of the prothorax, a thin patch of yellowish hairs in front of each eye; prothorax slightly convex, much broader than long, with two tufts of yellowish setose hairs on the disc, and six stout teeth on each side; scutellum very indistinct; elytra broad, convex, rounded at the side, the edges serrated, a small tuft of black hairs on each at the base, and a larger one common to both elytra behind and on the highest part of their convexity; antennæ about one-third the length of the whole insect, all the joints, except the last, furnished with three stiff setæ arising in the middle of each, two anterior and one posterior; palpi ferruginous; legs rough, with short thick hairs, tarsi ferruginous; eyes black; body beneath somewhat pitchy, coarsely punctured. Length 2¼ lines.

PHARAX [Colydiidæ].

Head short, transverse, rather widely dilated below the eyes, and deeply inserted in the prothorax. Antennæ short, eleven-jointed, the two basal incrassated, and nearly concealed above, the third longest, the rest gradually diminishing in length and becoming transverse, the last two forming a compact ovate club. Eyes small, round. Mentum rounded at the sides and in front. Terminal joint of the maxillary palpi triangular. Prothorax transverse, largely dilated and rounded at the sides, narrowed posteriorly, the disc very convex and irregular. Elytra connate, much broader than the prothorax at the base, short and irregular. Legs moderate; all the coxæ distant; femora robust; tibiæ fusiform, bordered externally with scale-like hairs; tarsi short, the basal joint longer than the second or third. Abdominal segments nearly equal.

This genus, in habit like *Ulonotus*, is allied to the last (*Hyberis*), from which the eleven-jointed antennæ and biarticulate club will at once distinguish it. The description of the mentum and palpi must be received with some hesitation, as they were examined *in situ*. The two specimens now before me are among those almost inexhaustible

I

captures of Mr. Fry at Rio, which perhaps, partly from their small size, and partly from the extremely limited area which many of the insects of that country affect, it is almost hopeless to expect can ever be obtained except by the most indefatigable and experienced collectors. The number of undescribed genera which are almost sure to be found in every extra-European collection that may be formed by an accomplished naturalist, should not be overlooked by those who are inclined to question the necessity of the multiplication of new names.

Pharax laticollis.　(Pl. VIII. fig. 1.)

P. ovatus, fuscus, tuberculiferus, grisco-setosus; antennis capitis latitudine æqualibus.

Hab. Brazil (Rio).

Ovate, dark brown, covered with short, stiff, scale-like hairs; head slightly concave above; prothorax somewhat bilobed anteriorly, its disc with four depressed tubercles; scutellum deeply set; elytra short, convex, with about ten tubercles on the disc, the posterior being the largest, the margins irregularly set with short stiff scales; antennæ, palpi, and tarsi ferruginous, the former about equal in length to the width of the head. Length 1½ line.

Chorites [Colydiidæ].

Head transverse, much narrower than the prothorax and deeply inserted in it, its supra-antennary borders slightly produced. Eyes large, and very rough, from the facets being prolonged into short spines. Antennæ short, slender, eleven-jointed, the first and second slightly incrassated, the third longest, the remainder to the ninth gradually decreasing in length, the tenth and eleventh forming an abrupt ovate club. Maxillary lobes ciliated, the external subtriangular, the internal narrower. Palpi short, claviform; the terminal joint of the maxillary ovate-cylindrical, of the labial ovate-oblong. Mentum subquadrate. Labium transverse, ciliated anteriorly. Prothorax very transverse, narrowed and sinuated anteriorly, as broad as the elytra at the base. Elytra convex, short, the sides gradually rounded to the apex. Legs small; coxæ, especially the posterior, very remote; femora compressed; tibiæ slightly enlarged at their extremity, ciliated externally, and terminated by two short spurs; tarsi short, slender, with long hairs beneath, the basal joint very distinct. Abdominal segments gradually decreasing in size.

The widely separated posterior coxæ narrow considerably the number of Colydian genera with which *Chorites* may be compared; at the same time, although the first abdominal segment is in every way larger than the others, there is not the decided difference we see in *Deratophrus*, *Sosylus*, &c.; and if we exclude these genera,

we are reduced to *Pycnomerus, Apeistus,* &c. To none of these,
however, is our insect closely related, the contiguity of the whole
base of the elytra to the prothorax completely isolating it from all
of them and their allies.

Chorites aspis. (Pl. VII. fig. 3.)

C. niger, subnitidus, squamis griseis indutus; antennis, palpis pedibusque
ferrugineis.

Hab. Borneo.

Broadly elliptical, black, rather glossy, covered with short erect pale
greyish scales, which are disposed in narrow rows on the elytra and
form a regular fringe round their margins and the sides of the prothorax; antennæ, palpi, and legs ferruginous, the tibiæ with a black stripe
externally and edged with a row of greyish scales; body beneath dull
black, thickly punctured, the throat only covered with yellow scales.
Length 2¼ lines.

There is a second species? in my collection, also from Borneo;
but, except in its much smaller size (about 1½ line long), and a few
black scales being interspersed among the others, there is little to
distinguish it.

Discoloma [Colydiidæ].

Erichson, Natur. der Ins. Deutschl. iii. p. 292.

Discoloma Fryi. (Pl. VII. fig. 2.)

D. piceo-ferruginea vel testacea, pubescens; elytris parce punctatis; antennis, palpis pedibusque dilutioribus.

Hab. Brazil (Rio).

Pitchy-ferruginous, in some specimens testaceous, sparingly pubescent; head rather closely punctured, inserted in a deep emargination
of the prothorax; prothorax very transverse, nearly twice as broad as
long, very finely punctured, the margins gradually but strongly dilated,
with its anterior angle rounded; scutellum small; elytra rather
broader than long, and as wide as the prothorax at the base, the disc
with several rather large, remote punctures, with a broad and strongly-
marked margin at the sides; antennæ, palpi, and legs pale ferruginous;
body beneath pitchy, with a few scattered hairs. Length 1¼ line.

Although Erichson has characterized *Discoloma* in very few words,
I cannot doubt that the insect described above is correctly referred
to that genus, as indeed Mr. Fry had previously suggested to me;
the only difficulty is, that *Discoloma* is said to have the basal joint
of its antennæ simple, or not enlarged, which is not the case in
the present species. However, the habit of the typical form appears
to agree with this, and is so remarkable—resembling some of the
Nitidulidæ (*Amphotis* for example)—whilst the structure so nearly

accords with *Cerylon*, in close proximity to which Erichson has placed the genus, that this discrepancy need not, for the present at least, necessitate the generic separation of the two insects. In addition to Erichson's description, the following generic characters (most of them the same as in *Cerylon*) may be noticed in *D. Fryi*:— Eyes narrow, transverse, scarcely prominent; external maxillary lobe long and very slender, ciliated at the apex (inner lobe not seen); maxillary palpi short, the first joint very small, the second greatly enlarged, the third subcylindrical, the fourth minute, aciculate; the labial palpi with the second joint enlarged, the third shortly conical; mandibles bidentate at their extremity; mentum small, quadrate; labium rounded anteriorly; tarsi very short, the three basal joints oblique, and hairy beneath.

GLYPTOLOPUS [Colydiidæ].

Erichson, Natur. der Ins. Deutschl. iii. p. 292.

Glyptolopus histeroides.   (Pl. VIII. fig. 5.)

G. late ovatus, piceus; prothorace elytrisque rugoso-costatis. *Hab.* Brazil (Rio).

Broadly ovate, pitchy-black; head coarsely punctured, small, vertical, scarcely visible above, narrowed below the eyes; antennæ twelve-jointed, the first large, incrassated, and uncovered at its insertion, the second short, not thicker than the third, the remainder becoming gradually stouter to the tenth and eleventh, the last small, closely enveloped in long silky hairs; prothorax semicircular, very convex, vaulted above and emarginate anteriorly, the centre with a broad longitudinal groove, and a stout interrupted costa on each side, the lateral margin strongly produced, the intervals coarsely punctured; scutellum triangular; elytra as broad as the prothorax at the base, but not continuous with it above, the sides rounded and gradually decreasing posteriorly, with five strong rugose costæ on each, the intervals coarsely punctato-granulate; all the coxæ distant, tibiæ fusiform, strongly fluted, not spurred, tarsi short; prosternum very strongly keeled, produced behind, and received in a notch of the mesosternum; first abdominal segment nearly as large as the rest together; body beneath coarsely punctured. Length 2 lines.

The few characters which Erichson has given of this genus, its very peculiar habit (resembling an *Onthophilus*), combined with the acicular palpi of the Ceryloninæ, and its habitat of Brazil, would seem to leave no doubt that the insect described above is correctly referred to *Glyptolopus*. The antennæ, however, are certainly twelve-jointed, while *Glyptolopus* is said to have only eleven. Has the little terminal joint been overlooked; and the ninth, which is nearly as large as the eleventh, been regarded as one of the three forming the club?

ALTHÆSIA [Mycetophagidæ].

Head deeply inserted in the prothorax, triangular, slightly dilated below the eyes. Antennæ longer than the prothorax, eleven-jointed, the last three forming an oblong perfoliate club. Eyes large, round, very prominent, rugose. Maxillary palpi with the second and third joints thickest, the terminal obconic, truncate; the labial short, triangular, approximate. Maxillary lobes narrow, nearly equal. Prothorax transverse, narrower and slightly emarginate in front, rounded at the side, the base bisinuated. Elytra slightly convex, margined, the base closely applied to the prothorax, but enlarging behind the shoulder, then rounded to the apex. Legs moderate; coxæ distant; tibiæ fringed externally, enlarging towards the extremity, and terminated by four or five short spines; tarsi slender, hairy beneath, four-jointed, the anterior with the penultimate very indistinct (male only?).

Resembles *Mycetophagus* in outline, but with a triarticulate club, and large round, very rugose and prominent eyes.

Althæsia pilosa.   (Pl. VI. fig. 4.)

A. piceo-brunnea, griseo-pubescens, pilosa; corpore infra pedibusque rufo-brunneis.

Hab. New Guinea (Dorey).

Pitchy-brown, covered with a close greyish pubescence combined with numerous soft, slender hairs; head scarcely half the breadth of the prothorax, sparingly punctured; prothorax with three grooves on each side, the inner two connected by a deep transverse one at the base: elytra slightly convex, widest behind the shoulder, with a very narrow margin; scutellum very small, triangular; body beneath and legs dark reddish-brown; abdomen, femora and tibiæ with a fulvous pubescence. Length 3 lines.

ATRACTOCERUS [Lymexylonidæ].

Palis. de Beauvois, Magaz. Encycl. 1802 (*sec.* Lacord.).

Atractocerus morio.   (Pl. VI. fig. 5.)

A. ater; elytris prothorace longioribus   alis chalybeatis; profemoribus coxisque testaceis.

Hab. Moluccas (Batchian).

Black; head nearly round, thickly punctured, closely covered with short erect black hairs; antennæ extending nearly to the end of the prothorax; eyes large, widely separated above; mandibles not projecting; prothorax narrower than the head, quadrate, hairy, shining; scutellum subtriangular, obtuse behind: elytra closely punctured, pubescent, nearly as long as the head and prothorax together; wings deep steel-blue, shining; abdomen black, slightly tinged with blue, with a very remote greyish pubescence; legs black, anterior coxæ and femora testaceous, the intermediate darker.   Length 11 lines.

DIOPTOMA [Lampyridæ].

Head exposed. Eyes very large, horizontally constricted, the upper portion smallest, the lower much larger, and completely contiguous. Antennæ short, claviform, subapproximate, deeply set on each side of the narrow prolongation of the front, twelve-jointed, the first two incrassated, the remainder forming an elongated club. Mandibles very slender, curved, not toothed. Palpi robust. Prothorax transverse, semicircular, not dilated at the sides. Scutellum rather large, triangular. Elytra as broad as the prothorax at the base, gradually rounded at the sides, narrow and flattened posteriorly. Winged. Legs moderate; intermediate coxæ not approximate; tarsi slender, the fourth joint not bilobed.

Although I do not hesitate to refer this most extraordinary insect to the Lampyridæ, yet it must be confessed that it is a very aberrant form, and suggests no affinity with any Malacoderm genus that I am acquainted with.   Its head (composed, at least externally, almost entirely of eyes, which are constricted in the middle like an hourglass) is fully exposed; the narrow vertex descends behind the upper portion of the eye, and fills in the space behind and between the constriction, and is prolonged in front to terminate in the labrum, although, from the presence of numerous coarse hairs, the existence of this organ cannot be positively asserted.   The antennæ are very short, scarcely extending to the prothorax, and show no traces of being serrated.   I am indebted for the only specimen I have seen to Dr. Ernest Adams, of University College, after whom I have named it.   The abdomen of the specimen having been cut away, apparently to facilitate (?) the mounting, the number of its segments cannot be ascertained: the abdomen itself, however, appears to have been very small; the metasternum must have exceeded it in length as well as in breadth.

Dioptoma Adamsii.   (Pl. V. fig. 2.)

D. fusca, parce pilosa; scutello elytrisque pallide grisescentibus, his plaga elongata fusca humerali.

Hab. India (Dacca).

Dark brown, rather sparingly clothed with pale semi-erect hairs, especially on the ·prothorax; head coarsely punctured, mandibles reddish-brown, antennæ and palpi pale yellowish; prothorax thickly and coarsely punctured; scutellum and elytra very pale greyish, inclining to yellow, the latter irregularly punctured with several slightly-raised longitudinal lines and a dark-brown elongate patch at the shoulder; body beneath and legs pale greyish. Length 3½ lines.

Cotulades [Tenebrionidæ].

Head subquadrate, exserted, but not constricted behind. Eyes small, lateral, round. Antennæ submoniliform, short, thick, very hairy, the basal joint longest, the rest to the tenth subequal, very transverse, the eleventh smaller, truncate. Labrum small, rounded anteriorly and ciliated. Mentum subquadrate, produced at the sides. Labium transverse, rounded in front. Palpi short, clavate, terminal joint ovate. Prothorax subquadrate, wider anteriorly. Elytra ovate, convex. Legs short; all the tarsal joints, except the last, very short.

To this genus belongs the *Tagenia leucospila* of Mr. Hope; the head, however, not contracted behind into a neck, and other characters show that it is very distinct from *Tagenia* [*Stenosis*]; at the same time it is difficult to point out a nearer ally. In this and the following genus the intermediate legs appear to be without trochanters.

Cotulades fuscicularis. (Pl. VII. fig. 5.)

C. niger, rugoso-punctatus; elytris obsolete albo-fasciculatis.
Hab. Australia (Melbourne).

Dull brownish-black; head and prothorax covered with large, coarse, nearly confluent punctures, and sparingly furnished with stiff, decumbent, scaly hairs; elytra coarsely striato-punctate, each with three indistinct ridges and with eight to ten short fascicles of brownish-white hairs, indeterminately arranged, but sometimes nearly wanting (from abrasion?); claws pale ferruginous; body beneath strongly punctured. Length 3 lines.

Elascus [Tenebrionidæ].

Head rather elongate, scarcely exserted. Eyes small, lateral, undivided. Antennæ short, hairy, eleven-jointed, the first longest, the rest transverse and more or less equal, except that the last is smaller than the preceding one. Palpi moderate, filiform, the terminal joint ovate, subacuminate. Mentum transverse, the angles rounded. Labium small, transverse. Prothorax subquadrate, irregular, much broader than the head, projecting in front, and lobed posteriorly, slightly dilated and serrated at the sides. Scutellum very small, quadrate. Elytra nearly parallel, broader than the prothorax. Legs short; femora and tibiæ compressed, the latter ciliated externally; tarsi very short and slender, the last joint nearly as long as the rest together.

This genus is not very far removed from the last; and, judging both from the figure and the description, I think that it is also allied to Erichson's *Latometus**.

Elascus crassicornis.   (Pl. VII. fig. 7.)

E. subdepressus, fuscescenti-varius ; antennis medio abrupte incrassatis.
Hab. Australia (Melbourne).

Rather broadly depressed, covered with coarse, curly, dusky-brown hairs varied with paler or greyish markings; head and prothorax greyish-brown, the latter with four tubercles on its disc and the projecting anterior portion strongly bilobed ; elytra bordered with hooked hairs, with three waved costæ on each, terminating posteriorly in as many tubercles, between which and the apex is another and larger one, a small oblique stripe behind the shoulder and a broad band near the apex ; antennæ greyish-brown, the terminal half darker, with the third joint much thicker than the two preceding, the fourth and succeeding joints gradually diminishing in thickness ; legs dark brown ; body beneath pitchy, with yellowish-brown scaly hairs.   Length 3 lines.

I have only seen two specimens, both of which were taken by Mr. Bakewell, at Melbourne, under the bark of trees composing a stock-yard fence.

Elascus lunatus.   (Pl. VII. fig. 8.)

E. subangustatus, fuscus, nigro-varius; elytris albo-fasciatis.
Hab. Australia (Melbourne).

Rather narrow, slightly depressed, covered with coarse scaly hairs, which are yellowish-grey on the head, but considerably darker on the prothorax and elytra, or nearly black, the latter having three whitish bands (the two anterior crescent-shaped, but sometimes nearly coalescing, the posterior straight) ; prothorax with four tubercles on its disc, the anterior projecting portion rather broadly bilobed, each lobe forming (so to speak) an additional tubercle ; elytra coarsely seriato-punctate, each with three costæ, the inner nearly obsolete except at the base ; antennæ not abruptly thickened in the middle, yellowish varied with dark brown, especially the three terminal joints ; legs ferruginous, more or less marked with dark brown ; body beneath covered with greyish-yellow scaly hairs.   Length 2½ lines.

The post-prothoracic lobe is less developed in this species than in the former, or, in other words, it is broader and less abruptly defined.   The two specimens (also captured by Mr. Bakewell) now before me differ considerably in depth of colour and amount of white on the elytra ; but in this, as in other instances, the pattern is the same.

* Wiegmann's Archiv, 1842. p. 213. pl. 5. fig. 3.

Docalis [Tenebrionidæ].

Head rounded, exserted, the antennary orbit nearly dividing the eye.
Antennæ short, covered with numerous small flattish hairs, the first
three joints longest, the rest transverse, the tenth larger than the
eleventh. Mandibles stout, bifid at the apex. Palpi robust, terminal joint
of the maxillary short, stout, of the labial obconic, obtuse; external
maxillary lobe short, triangular, fringed, the inner narrow, toothed.
Mentum arising within the jugular plate. Prothorax subquadrate,
scarcely wider than the head. Elytra ovate-oblong, broader than the
prothorax. Legs short, the intermediate furnished with trochanters;
coxæ not contiguous; tibiæ not spurred; tarsi with all the joints except the last very short and fringed with spiny hairs. Prosternal process quadrate. Mesosternum depressed.

The *Tagenia funerosa* of the Rev. F. W. Hope is, I think, referable to this genus; and, trusting solely to recollection of his type,
now in the Taylor Institute at Oxford, it is very close to, if not
identical with, my *D. degener*; but without certainty on this point,
it is better to assume that they are distinct. The genus seems to
be referable to the Scaurinæ, and, so far as my knowledge of the
group extends at present, it might follow *Ammophorus*. The structure of the mouth, in reference to what I have called the "jugular
plate," but which appears to be the "*sous-menton*" of M. Lacordaire,
is very similar, judging from that author's description, to that of
Nyctoporis, which genus immediately precedes *Ammophorus*. The
larger penultimate joint of the antennæ is suggestive in a slight
degree of the club of many Colydian genera; indeed, there are so
many points of resemblance between several of the Heteromera and
the Colydiidæ, as to justify a doubt whether they may not be more
than mere analogies.

Docalis exoletus. (Pl. VIII. fig. 9.)

D. oblongo-ovatus, fuscus; prothorace transverso.
Hab. Australia (Melbourne); Tasmania.

Oblong-ovate, dark brown, everywhere covered, but not very closely,
with semi-erect, stiff black scales (hairs), intermixed, especially on
the head and prothorax, with rusty-white; prothorax slightly broader
than long; scutellum rounded behind; elytra coarsely seriato-punctate,
marked with several slightly elevated longitudinal lines, which are
severally crested with a row of whitish scales; body beneath punctured,
each puncture enclosing a short rusty hair. Length 2 to 3 lines.

For my knowledge of this and the species of the two preceding
genera, I am indebted to Robert Bakewell, Esq., who informs me
that they, and many other insects as well, are found beneath the

bark of logs which are piled one on another in the formation of stockades. Few of the many collectors in Australia appear to be aware of the novelties which a careful examination of such localities would afford them.

Docalis degener.

D. oblongo-ovatus, præcedenti angustior, niger; prothorace æquali.
Hab. Tasmania.

Narrower and darker than the last, with the prothorax at least as long as it is broad, the scales whiter and less numerous and the punctures larger, and the longitudinal lines on the elytra more prominent. Length 2 lines.

SPHARGERIS [Tenebrionidæ].

Head small, transverse, abruptly contracted below the eyes. Antennæ eleven-jointed, very short, gradually increasing in thickness from the third, which is longest, the second minute, the first incrassated. Eyes lateral, very small, round. Labrum narrow, not covering the mandibles, which are bifid at the tip. Maxillary lobes narrow, the terminal joint of their palpi subsecuriform. Mentum subcordate, narrower behind. Labium bilobed and ciliated anteriorly; labial palpi long, the terminal joint ovate, pointed. Prothorax short, transverse, narrower anteriorly, rounded at the sides. Elytra shortly ovate, very convex. Legs short, more or less covered with spinous hairs; tibiæ triangular, strongly spurred, the anterior sinuated externally; tarsi short, the basal joint longer than the second. Prosternum compressed, cariniform.

Closely allied to Mr. White's genus *Chærodes* (Voyage of the Erebus and Terror, Ins. p. 12. tab. 2. fig. 12), but differs essentially in the antennæ, *Chærodes* having (*inter alia*) a triarticulate club (*see* Pl. V. fig. 10); in both, however, they are eleven-jointed.

Sphargeris physodes.   (Pl. V. fig. 9.)

S. testaceus, subnitidus, punctulatus; oculis mandibulisque nigris.
Hab. Australia (Melbourne and Adelaide).

Broadly ovate, very convex, smooth, shining, testaceous, closely and finely punctured; scutellum small, triangular; antennæ about as long as half the breadth of the head; eyes and mandibles black; body beneath darker, punctured, with short scattered hairs. Length 3 lines.

CHÆTYLLUS [Tenebrionidæ].

Head subtriangular, rounded posteriorly, larger than the prothorax, its supra-antennary borders forming a short, thick, elevated protuberance. Antennæ moderately long, eleven-jointed, the first incrassated, the second minute, the third longest, the rest more or less moniliform and becoming gradually thicker upwards. Eyes lateral, small, round. Maxillary palpi strongly securiform, the labial very short and thick. Prothorax narrower than the head, much contracted behind. Scutellum none. Elytra connate, very convex, broadly elliptical. Legs moderate; anterior coxæ globose, not contiguous; tibiæ unarmed, hairy at the base internally; tarsi short, thick, hairy beneath, the basal joint longer than the second, the penultimate bilobed. Prosternum produced, rounded posteriorly, and remote from the mesosternum.

An examination of the mouth might throw some light on the affinities of this very curious little insect; but as the only specimen I have seen belongs to the British Museum, and moreover is not in very good condition, this cannot be done at present. In habit it resembles the Anthicidæ, but the globose anterior coxæ separate it from that family; the bilobed tarsi, an unusual character amongst the Tenebrionidæ, suggest an analogy, or perhaps an affinity, with *Phymatodes* and *Phobelius*. It is one of the many important captures of Mr. Bates in the valley of the Amazons; and as that gentleman is preparing a series of papers on some of the insects of his extensive collections, it is to be hoped that this and many other curious forms which he possesses will be at no distant date more amply illustrated.

Chætyllus anthicoides. (Pl. VI. fig. 8.)

C. niger, nitidus; prothoraco elytrisque tuberculatis, tuberculis setigeris; tarsis pallidioribus.
Hab. Brazil (Ega).

Black, shining; head coarsely punctured, with scattered, erect, setulose hairs, a semicircular groove between the antennary orbits; prothorax and elytra covered with large tubercular elevations, arranged in rows on the latter, each of which bears a long, erect, setose hair; tarsi and base of the tibiæ internally with pale silky hairs; labial and maxillary palpi at the base pale ferruginous; antennæ setigerous, as long as the head and prothorax together. Length 2 lines.

DIPSACONIA [Tenebrionidæ].

Head small, rather narrow and elongate below the eyes, deeply inserted in the prothorax. Eyes transverse, undivided. Antennæ rather short, submoniliform, slightly hairy, the basal joint incrassated, the second very short, the third longest, the remainder gradually decreasing in length, but becoming broader and transverse, to the ninth and tenth, the eleventh subovate. Labrum rounded anteriorly. Maxillary palpi rather long, claviform, the last joint large, ovate, truncate; the labial very small; external maxillary lobe broad, strongly ciliated. Mentum

quadrate. Labium very transverse. Prothorax narrower than the
elytra, transverse, sinuated anteriorly, its surface regular. Elytra
rather long, slightly rounded at the sides. Legs moderate; tibiæ bicalcarate, ciliated externally; tarsi slender.

Allied to *Ulodes*, Er., which differs in the following points. In
Ulodes the head is short, not being prolonged below the eyes; the
joints of the antennæ are subequal and transverse, surrounded by a
dense whorl of squamose hairs; the surface of the prothorax is very
irregular; the elytra are short, and the body generally is covered
with short crisp scales. To *Ulodes* I refer *Bolitophagus Saphira*,
Newm., and *Endophlœus variicornis*, Hope. My genus *Byrsax* (*ante*,
p. 42) is also a member of this group of Tenebrionidæ (Bolitopha-
ginæ): it is true I cannot quite satisfy myself that it is heteromerous, but I have no doubt a minute basal joint exists; and in
other respects it appears to be congeneric with *Diaperis horrida*, Ol.
(*Asida horrida*, Walk.). *Trox cornutus*, Fab., is also referable to
Byrsax.

Dipsaconia Bakewellii.   (Pl. VII. fig. 6.)

D. elliptico-ovata, pilosa, fulvo-brunnea; elytris nigro-variegatis.
Hab. Australia (Melbourne).

Elliptic-ovate, brownish-fulvous, covered with short decumbent
hairs, among which others longer, nearly erect and slightly curved, are
interspersed; prothorax nearly as wide as the elytra at the base; scutellum rather indistinct, subtriangular; elytra nearly parallel at the
sides, rounded at the apex, striato-punctate, each with three costæ, and
varied with four or five dull-black band-like marks; antennæ brown;
body beneath ferruginous-brown, very sparingly pubescent. Length 3½
lines.

In this and the following species, both of which we owe to Mr.
Bakewell's researches, may be noticed, in certain lights, a glowing
fiery-red tubercle at the bottom of each elytral puncture.

Dipsaconia pyritosa.

D. elongato-ovata, hirta, rufo-fusca; prothorace elytrisque nigro-variegatis.
Hab. Australia (Melbourne).

Elongate-ovate, reddish-brown, closely covered with short, thick,
strongly hooked hairs; prothorax narrower than the elytra at the base,
the disc with a large irregular blackish patch; scutellum indistinct,
subquadrate; elytra rather broader behind, striato-punctate, marked
with several irregular, dull brownish-black patches; antennæ brown;
body beneath and legs ferruginous-brown, sparingly pubescent. Length
3½ lines.

Tithassa [Tenebrionidæ].

Head small, exserted, its anterior border incrassated. Antennæ stout, moderately long, the first and second joints scarcely thicker than the third, which is longer, the remainder to the eighth short, the last three forming an oblong, loose, compressed club. Eyes small, lateral, round. Epistome and labrum narrow, not covering the mandibles, the latter broadly emarginate. Mandibles bifid at the apex; terminal joint of the palpi ovate, subacuminate, the second joint of the labial larger than the third; maxillary lobes subequal, fringed. Mentum subquadrate. Labium rounded. Prothorax transversely subquadrate, narrower than the elytra, its margins dilated. Elytra large, convex, broadly ovate. Legs small; coxæ not approximate, the anterior cylindrical, transverse; tibiæ not spurred; tarsi pubescent beneath, the penultimate joint dilated. Prosternum pointed behind; mesosternum depressed; post-intercoxal process triangular.

The majority of the characters of this genus point, as it appears to me, to the Diaperinæ, but the differently-formed tarsi and the disproportion between the prothorax and elytra forbid its union with that group. At the same time, the antennæ come nearer those of *Pentaphyllus* "in plan" than any other heteromerous genus that I am acquainted with. It seems to be a common Rio insect.

Tithassa corynomelas. (Pl. V. fig. 7.)

T. testaceo-lutea, nitida, punctata; oculis, antennisque, ab articulo sexto, nigris.

Hab. Brazil (Rio).

Dark glossy testaceous, or luteous-brown, irregularly punctured above, with a few very fine and extremely scattered slender hairs; eyes and last five joints of the antennæ, including a portion of the sixth, which are also more hairy than the rest, black. Length 3 lines.

Chariotheca [Helopidæ].

(Dej.) Catal. des Coléopt.

Head moderate, subquadrate. Eyes large, transverse, contiguous to the prothorax. Antennæ short, claviform, the first joint nearly concealed above by the antennary orbits, the four or five terminal joints compressed and, except the last, more or less transverse. Labrum rounded anteriorly. Maxillary palpi with the last joint securiform, the labial ovate, truncate; maxillary lobes short, strongly ciliated. Mentum subquadrate. Labium slightly expanded at the sides, entire and ciliated in front. Prothorax transverse, nearly as broad as the elytra at the base, rounded at the sides, scarcely emarginate anteriorly. Elytra elongate, their greatest breadth behind the shoulders, slightly curved at the sides. Legs rather slender; tarsi hairy beneath, the basal joint longer than the succeeding one. Prosternum pointed behind, with a narrow impression in the middle; mesosternum notched for the reception of the prosternum; post-intercoxal process pointed anteriorly.

This unpublished genus of Dejean's was placed by him nearly at the end of his *Tenebrionites*, an heterogeneous assemblage, including as it does *Melandrya*, *Pytho*, *Pezodontus*, *Camaria*, &c. With the last of these genera, however, and with its allies, *Chariotheca* must be placed.

Chariotheca coruscans.   (Pl. VI. fig. 7.)

C. atra, nitida; elytris cyaneis; corpore infra, antennis pedibusque ferrugineis.

Hab. Moluccas (Batchian).

Deep black, smooth, shining; head and prothorax lightly and irregularly punctured; scutellum triangular; elytra rich indigo-blue, seriatopunctate (about nine rows), with numerous smaller punctures irregularly crowding the interstices; antennæ not longer than the breadth of the head, reddish-ferruginous, the last five joints with a few short scattered greyish hairs; palpi and legs, particularly the tibiæ and tarsi, reddish-ferruginous; body beneath ferruginous, inclining to chestnut. Length 4½ lines.

Chariotheca litigiosa.

C. atra, nitida; elytris chalybeo-cyaneis; antennis tarsisque ferrugineis; corpore infra, femoribus tibiisque atris.

Hab. New Guinea (Aru).

Deep black, smooth, shining; head with crowded oblong punctures, often three or four more or less confluent, and then forming short longitudinal folds in the spaces between them; prothorax with small scattered punctures; scutellum rather small, triangular; elytra dark green, punctured as the last; antennæ, palpi, and tarsi reddish-ferruginous; body beneath, femora and tibiæ black. Length 4½ lines.

Rather narrower than the former, the scutellum smaller, the head differently punctured, the colour less brilliant, &c.

Chariotheca cupripennis.

C. atra, nitida; elytris cupreis; corpore infra, antennis pedibusque piceis.

Hab. New Guinea (Dorey).

Deep black, shining; head, especially between the eyes, with many oblong punctures; prothorax irregularly punctured; elytra seriatopunctate, the interstices crowded with very minute punctures, copperred, the suture rich green; antennæ and palpi ferruginous-brown; body beneath and legs pitchy. Length 4 lines.

OMOLIPUS [Helopidæ].

Head transverse, vertical, sulcated in front. Antennæ short, gradually increasing in thickness, the two basal joints small, the third longest, the fourth to the seventh obconical and decreasing in length, the last four submoniliform, compressed. Eyes transverse, partially divided in front. Labrum rounded anteriorly and ciliated. Mandibles bidentate at the apex. Maxillary palpi securiform; the labial approximate at the base, with the terminal joint triangular. Maxillary lobes small, the inner strongly hooked. Labium transverse. Mentum subtriangular, truncate at the base, carinated in the middle. Prothorax convex, rounded in front and at the sides, closely applied to the elytra, its parapleuræ distinct. Scutellum small, triangular. Elytra connate, ovate, convex. No wings. Legs stout; anterior coxæ globular, not contiguous; tibiæ straight, unarmed; tarsi short, all the joints except the last dilated. Prosternum wedge-shaped, produced, with a deep central impression; mesosternum notched for the reception of the prosternum.

In characterizing *Œdemutes* (*ante*, p. 51), the semilunar, sulcated anterior portion of the head was described as the epistome, and M. Lacordaire appears to have done the same in his description of *Sphærotus**. The real epistome, however, is inserted *beneath* the anterior border, and in *Sphærotus curvipes* is completely hidden by it; but, on the other hand, it is almost entirely exposed in another common species, *Sphærotus gravidus*. In *Omolipus* (at least in the species described below; for the character scarcely seems to be of generic value), the labrum, which is rather strongly developed, also appears to be inserted directly beneath the anterior border of the head, and the epistome is therefore not apparent. The nearest affinity of *Omolipus* is probably *Misolampus*, from which, among other characters, the presence of a very distinct scutellum will at once distinguish it. This genus is another exception to the absence of the hook on the internal maxillary lobe, a character which at one time was supposed to distinguish the Helopidæ from the Tenebrionidæ. Another exceptional character is the approximation of the base of the labial palpi, which are inserted in front of the broadly transverse, membranous lower lip.

Omolipus corvus. (Pl. VI. fig. 9.)

O. ater, nitidus; elytris punctato-impressis; antennis tarsisque pallidioribus.

Hab. Australia (Melbourne).

Deep glossy black; head and prothorax very minutely punctured; elytra narrower than the prothorax, each with about nine rows of deeply

* Gen. des Coléopt. v. p. 446.

impressed punctures; legs smooth and shining, tarsi brownish; antennæ shorter than the prothorax, paler at the apex; body smooth beneath. Length 5–6 lines.

Rhinosimus [Salpingidæ].
Latreille, Gen. Crust. et Ins. ii. p. 231.

Rhinosimus Wallacei.

R. atro-chalybeus, nitidus; rostro pedibusque rufis; elytris purpureis; antennarum funiculo tarsisque luteis.

Hab. New Guinea (Dorey).

Ovate, slightly depressed, finely punctured, smooth and shining; head deep steel-blue, the rostrum dark reddish-yellow, rather dilated at the apex, the antennæ inserted at about the middle, the last three joints, forming a strongly marked club, black; prothorax deep steel-blue, narrower than the elytra; scutellum very transverse; elytra dark purple; femora and tibiæ yellowish-red, tarsi pale brownish-yellow; body beneath chestnut-brown. Length 2½ lines.

Zonitis [Cantharidæ].
Fabricius, Syst. Entom. p. 126.

Zonitis Downesii.

Z. breviusculus, luteus, punctulatus; antennis, basi excepta, nigris; tarsorum articulo ultimo apiceque elytrorum infuscatis.

Hab. India (Bombay).

Rather short, brownish-yellow, the upper surface minutely punctured; head and prothorax rather glossy, and together considerably more than half the length of the elytra; scutellum rounded posteriorly; elytra much wider than the prothorax at the base, the apex clouded with brown; antennæ scarcely extending to the base of the prothorax, black, the two basal joints yellow; palpi and mandibles at their tips, and the last joint of all the tarsi above and their claws (more or less) dark brown; legs covered with short silky hairs. Length 6 lines.

Dedicated to Ezra Downes, Esq., of Calcutta, who, during his residence at Bombay, collected and sent to this country many interesting insects from that locality, and after whom was named, as its discoverer, the very fine and remarkable Prionian *Cantharocnemis Downesii.*

Trigonops [Curculionidæ].
Guérin-Méneville, Rev. Zool. 1841, p. 128.

Trigonops Jekelii.   (Pl. VII. fig. 9.)

T. piceus, punctato-granulatus, squamis viridescentibus tectus; elytris brevibus, perpendiculariter deflexis; femoribus basi rufis.

Hab. Celebes (Manado).

♂ Elytris convexis, angulis posticis cornutis.
♀ Elytris deplanatis, angulis posticis muticis.

Ovate, dark pitchy-brown, sparingly furnished above with pale yellow-ish-green scales; rostrum longer than the head, gibbous below the eyes, and separated from them by a semicircular depression, with a broad longitudinal furrow in the middle; prothorax shortly ovate, closely granulated, and covered with coarse deep punctures; scutellum none; elytra very short, perpendicularly bent down behind, roughly punctato-granulated, slightly convex in the male, with the posterior angle produced into a long flexible process, flat and depressed in the female, and without any prolongation; legs moderate, furnished with stiff scattered hairs, the femora orange-red, except at the apex (in the female darker); antennæ black, shorter than the body, slightly hairy; body beneath pitchy, coarsely punctured. Length 3½ lines (♂). 3 lines (♀).

BLAPSILON [Cerambycidæ].

Head short, scarcely convex in front. Eyes small, lateral, deeply emargi-nate. Antennæ shorter than the body, sublinear, distant at the base, the first joint thickened, shorter than the third, which is longest, the fourth moderate, the remainder very short and subequal. Labrum small, slightly emarginate. Mandibles robust. Palpi stout, the terminal joint elongate-ovate, truncate. Mentum very short and transverse. Pro-thorax broader than long, narrower in front. Scutellum elongate, pro-duced anteriorly. Elytra ovate, broader than the prothorax at the base, elevated in the middle, and produced at the shoulder into a short, hooked, horizontal process. Legs moderate; coxæ distant; tarsi short, very slightly dilated. Prosternum received into a notch of the mesosternum.

The scutellum of this genus is remarkable. It is not only un-usually narrow and somewhat hexagonal in form, but it is projected forwards on the prothorax, which is probably notched for its recep-tion, although this point cannot be ascertained without risk of in-jury to the specimen. *Blapsilon* must be placed near *Tmesisternus*.

Blapsilon irroratum. (Pl. V. fig. 8.)

B. fusco-piceum, maculis hirtis ochraceis punctisque impressis adspersis. *Hab.* New Caledonia.

Broadly ovate, dark pitchy-brown, the whole upper surface, except the scutellum, covered with small, round, hairy ochraceous spots and deeply impressed closely-set punctures; body beneath pitchy-brown; anterior tibiæ and tarsi paler. Length 7 lines.

There are two specimens in the British Museum, collected during the surveying expedition of H.M.S. Herald.

AUXA [Lamiidæ].

Head small, convex in front, the vertex elevated. Antennæ setaceous, longer than the body, pedunculate, the first joint thickened, pyriform,

K

the third longest, slightly curved, the rest subequal. Eyes small, deeply divided. Epistome and labrum large and transverse, the latter broadly emarginate. Palpi long, acuminated. Prothorax elongate-ovate, broader than the head, very irregular, toothed at the sides. Elytra narrow, convex, tapering posteriorly. Winged. Legs stout; femora clavate; tarsi short. Prosternum dilated posteriorly; mesosternum slightly bilobed.

The unusually large prothorax of this insect and its narrow, tapering elytra at once suggest some *Dorcadion* form, but its real position appears to be with *Pogonocherus* and its allies. The specimen from which the description has been drawn up is in the Hopean collection at Oxford.

Anxa amplicollis.   (Pl. VI. fig. 2.)

A. fuscata, subtilissime pubescens; elytris pallidioribus, plagis magnis duabus, una basali, alteraque apicali, albescentibus.

Hab. Madagascar.

Dull brown, finely pubescent; prothorax very irregular, transversely corrugated, the centre armed with two strong recurved teeth and a shorter tooth at the side; scutellum very transverse, whitish; elytra narrow, apiculate, spined at the shoulder, pale brown, a large whitish irregular patch at the base and another at the apex; antennæ rather longer than the body, ferruginous-brown, slightly ciliated beneath; palpi testaceous; legs dark brown, rather glossy, the base of the femora paler, a whitish patch on the posterior; body beneath with a greyish-white pubescence.   Length 3½ lines.

Cacia [Lamiidæ].

Newman, The Entom. p. 290.

Cacia anthribioides.   (Pl. V. fig. 5.)

C. atra, pubescens; capite prothoraceque strigis, elytrisque (parte antica) albo-cinereis; antennis tarsisque albo-annulatis.

Hab. Amboyna.

Deep black, covered with a very short dense pubescence; head below the eyes, and two nearly confluent stripes between them, ashy-white, lip margined with white; prothorax longer than wide, subcylindrical, a little bulging at the sides, with a broad central stripe and the sides ashy-white; scutellum subquadrate, the apex white; elytra much wider than the thorax at the base, rather short, very slightly receding towards the apex, which is rounded, with considerably more than its basal half white, except at the shoulders and around the scutellum ashy-white, a few white spots also at the apex; legs rather short and robust, slightly tinged with ashy, the two basal joints of all the tarsi white; antennæ nearly twice as long as the body, the base of the third, fourth and fifth joints white, the fourth with a slight tuft of hairs at its apex; body beneath ashy.   Length 8 lines.

OMOSAROTES [Lamiidæ].

Head exserted, vertical, quadrate in front. Eyes very deeply divided, the two portions connected only by a narrow line. Antennæ distant, robust, shorter than the body, pedunculate, and ciliated beneath, the first joint slightly incrassated, the third longest, the rest gradually decreasing in length. Epistome very short. Labrum small, transverse, rounded. Palpi slender, subacuminate. Prothorax arched, **narrower** than the elytra, rounded in the middle, contracted anteriorly and posteriorly, the sides strongly toothed. Scutellum quadrate. Elytra short, narrow, broadest at the base, convex. Legs moderate; tibiæ compressed, the anterior emarginate internally; tarsi very short, the basal joint triangular. Prosternum broad, rounded posteriorly; mesosternum sub-bilobed.

This genus, with *Neopadus*, appears to enter into a small group of South American Longicorns, of which the *Cerambyx sericeus* of Perty may be considered as the type. This is one of Mr. Bates's rarest captures, he having never met with more than two specimens ; one is now in my collection, the other in his own.

Omosarotes singularis. (Pl. VIII. fig. 5.)

O. atro-piceus, crinitus, pube sparsa griseo-fulva varius ; elytris basi pedunculo-fasciculatis.

Hab. Brazil (Para).

Pitchy-black, with long slender scattered hairs, particularly on the posterior part of the elytra and legs, and rather thinly covered with a greyish-yellow pubescence, which is most predominant on the prothorax and basal half of the elytra, forming also a sort of band, which is margined with a little white anteriorly, across their posterior third ; head narrower above the eyes, the peduncles bearing the antennæ rather distant, with a longitudinal groove between them ; lateral tooth of the prothorax on the middle ; a sharp carina half the length of the elytra terminating at the humeral angle, the side below it bent abruptly down, near the base an elevated protuberance bearing a fascicle of long, nearly erect black hairs ; tibiæ with a line of thickly-set yellowish hairs externally ; body beneath deep black, the throat, breast and abdomen very glossy. Length 5 lines.

LANGURIA [Languriidæ].

Latreille, Gen. Crust. et Insect. iii. p. 65.

Languria illætabilis. (Pl. V. fig. 4.)

L. elongata, rubro-fusca ; elytris chalybeo-viridibus ; antennarum clava, pedibusque fuscis.

Hab. Natal.

Narrowly elongate, dark reddish-brown, smooth, shining ; head and prothorax finely punctured, the latter much narrower posteriorly ; scutellum subcordate, reddish-brown ; elytra narrow, parallel, striato-punctate, dark steel-green ; antennæ pale at the base, the club black ;

legs dark brown; eyes black; body beneath smooth, glossy black, the breast reddish-brown. Length 3 lines.

Languria pulchella.

L. elongata, fulva; prothorace medio sulcato; capite elytrisque viridibus; antennarum clava fusca; pedibus flavis.

Hab. Natal.

Narrowly elongate, smooth, shining; head dark green; prothorax finely punctured, reddish-yellow, longitudinally grooved in the middle: scutellum subcordate, black; elytra punctato-striate, glossy bluish-green; antennæ dark brown, paler at the base; legs yellow; body beneath glossy black, the breast reddish-yellow. Length 3 lines.

This and the above are probably distinct from the true *Languriæ*.

EXPLANATION OF THE PLATES.

PLATE V.

Fig.
1. *Acropis aspera.* Para.
2. *Dioptoma Adamsii.* Dacca.
3. *Bothrideres succineus.* Rio.
4. *Languria illætabilis.* Natal.
5. *Cacia anthriboides.* Borneo.
6. *Prostomis morsitans.* Darjeeling.

Fig.
7. *Tithassa corynomelas.* Rio.
8. *Blapsilon irroratum.* Lord Howe's Island.
9. *Sphargeris physodes.* Melbourne.
10. Antenna of *Chærodes trachyscelides,* White.

PLATE VI.

1. *Sosylus sulcatus.* Para.
2. *Aura amplicollis.* Madagascar.
3. *Asprotera inculta.* Natal.
4. *Althæsia pilosa.* New Guinea.
5. *Atractocerus morio.* Moluccas.

6. *Dustarcus confinis.* New Guinea.
7. *Chariotheca coruscans.* Moluccas.
8. *Chætyllus anthicoides.* Ega.
9. *Omolipus corvus.* Moreton Bay.

PLATE VII.

1. *Hyberis araneiformis.* Borneo.
2. *Discoloma Fryi.* Rio.
3. *Chorites aspis.* Borneo.
4. *Rhyssopera areolata.* Tasmania. (Trophi of *R. illota.*)
5. *Cotulades fascicularis.* Melbourne.

6. *Dipsaconia Bakewellii.* Melbourne.
7. *Elascus crassicornis.* Melbourne.
8. *Elascus lunatus.* Melbourne.
9. *Trigonops Jekelii.* Celebes.

PLATE VIII.

1. *Pharax laticollis.* Rio.
2. *Glyptolopus histeroides.* Rio.
3. *Lemmis cælatus.* Rio.
4. *Distaphyla mammillaris.* Para.
5. *Omosarotes singularis.* Para.
6. *Ethelema luctuosa.* Rio.

7. *Docalis exoletus.* Melbourne.
8. *Anarmostes sculptilis.* Rio.
9. *Glæania ulomoides.* Rio.
9 a. Its anterior tarsus seen from beneath.

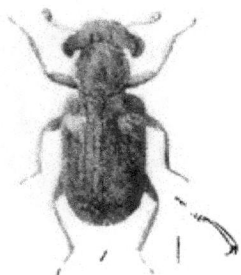

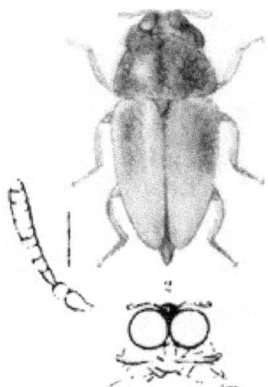

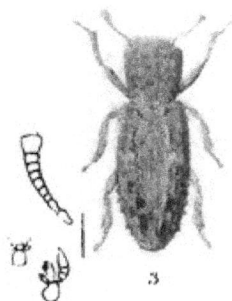

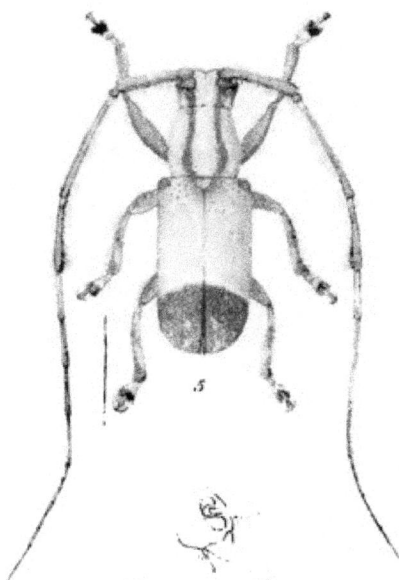

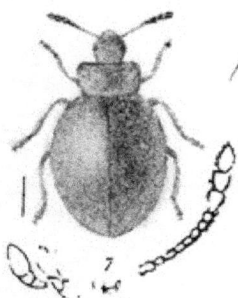

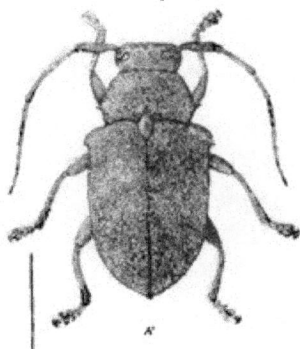

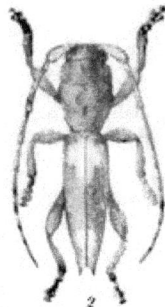

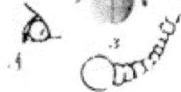

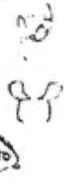

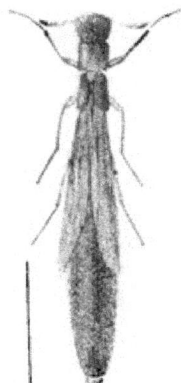

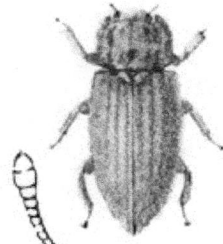

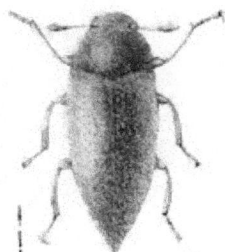

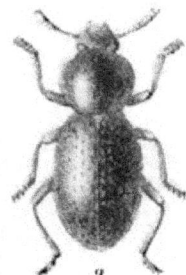

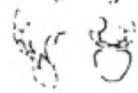

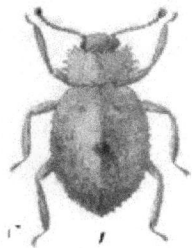

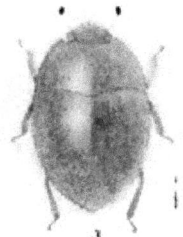

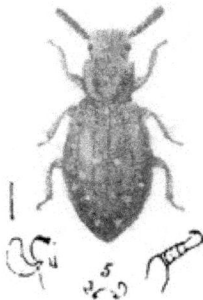

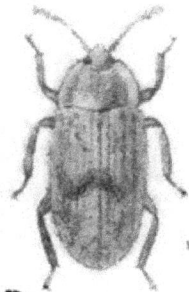

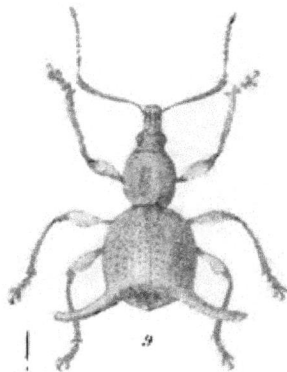

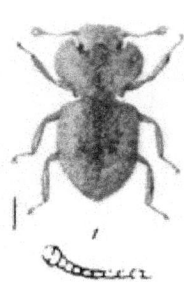
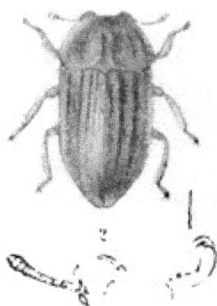
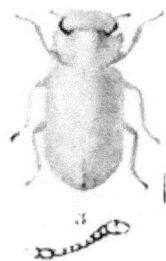

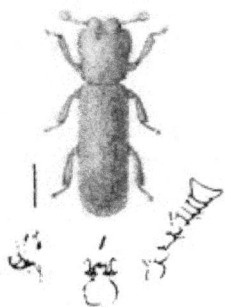
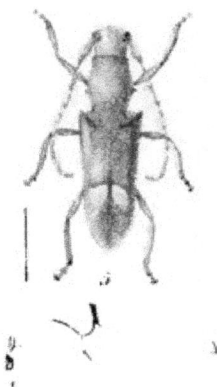
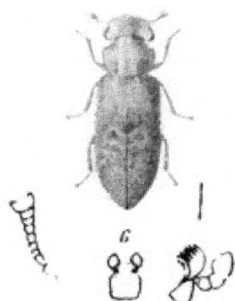

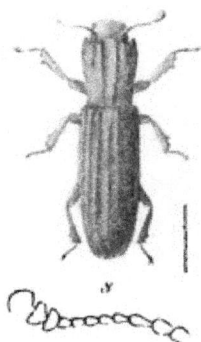
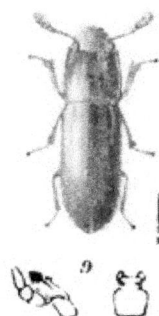

P lith

[*From the* JOURNAL OF ENTOMOLOGY *for* MAY 1862.]

Notices of new or little-known Genera and Species of Coleoptera.

By FRANCIS P. PASCOE, F.L.S., &c.

[Continued from p. 132.]

PART III.

MELAMBIA [Trogositidæ].

Erichson, in Germar, Zeitsch. v. p. 451.

Melambia maura.

M. elongata, atra; prothorace vix trausverso, lateribus basiu versus rotundatis.

Hab. South Africa (N'Gami).

Elongate, black; head dull black, closely covered with oblong punctures having the appearance of a small granule in the centre of each, mandibles also covered with oblong punctures except at the bifid apex; antennæ as long as the breadth of the head behind, the first joint punctured, the rest glabrous with a few hairs only on the club: prothorax shining black with the anterior angles obtuse, the sides rounded rapidly to the base, the posterior angle nearly obsolete, covered with oblong punctures, those at the side only granulated; scutellum transverse, with 6-8 punctures in two rows; elytra dull black, seriate-punctate, the punctures coarse, oblong, and in double lines, the intervals smooth, and slightly elevated; femora and tibiæ simply punctured; body beneath pitchy black with granulated punctures. Length 7 lines.

Melambia memnonia.

M. subelongata, atra; prothorace transverso, disco subplanato, antice incrassato, basi lata, angulis posticis acutis; elytris obscure fuscis.

Hab. Ceylon.

Subelongate, black; head covered with rather closely set, oblong, granulated punctures, mandibles with small simple punctures extending to the bifid apex; prothorax black, slightly shining, punctured as on the head, but less closely, and the punctures with granulated bases confined to the sides, anterior margin thickened immediately above the vertex, the disk flattened behind the thickened parts, side slightly rounded, then shortly curving inwards, and terminating at a sharp angle in a broad base; scutellum transverse, with eight or ten scattered punctures; elytra opake, nearly black, with a slight chestnut-brown tinge, punctured in double rows, the outer row with its punctures about a third or a fourth of the size of the inner, which latter are more or less impressed on the side of the raised lines between the rows; legs pitchy, the femora and tibiæ punctured; body beneath, under side of the mandibles, and palpi reddish-pitchy, the former with scattered punctures, each nearly entirely occupied by a smooth granule. Length 6 lines.

In the form of the prothorax this species approaches *M. gigas*, Fab., and apparently also *M. striata*, Or., both from Senegal; but the former is larger and more robust, with bluish-black elytra, &c., and the second is distinguished by its more punctured and remarkably transverse scutellum, &c. *M. crenicollis*, Guér., from India, seems to be a smaller species with a differently shaped prothorax, with its sides sufficiently crenated to suggest the specific name.

Melambia funebris.

M. subelongata, obscure atra; prothorace transverso, disco leviter convexo, basi sublata, angulis posticis acutis.

Hab. Cambodia.

Very like the last, but differs in the following particulars: prothorax longer, more rounded at the sides, and more contracted at the base, slightly but regularly concave over the whole disk, the anterior margin not in the least thickened ; elytra with the lines between each double row of punctures more raised, the punctures (more nearly equal in size) and the lines themselves gradually disappearing towards the shoulder; colour a dull black, without any tinge of brown.

It is quite possible that this may be only a local variety ; but, with the members of a genus so closely allied as they are in *Melambia*, this cannot be assumed until we obtain intermediate forms.

BRONTES [Cucujidæ].

Fabricius, Syst. Eleuth. ii. p. 97.

Brontes lucius.

B. ferrugineus, setulosus ; prothorace lateribus denticulatis, dente antico incrassato ; elytris striato-punctatis, marginibus infuscatis.
Hab. Sydney.

Ferruginous brown, covered with short, dark, setulose hairs ; head rather exserted, the vertex somewhat depressed ; eyes dark brown ; antennæ longer than the body, with a slight greyish pubescence, the first joint nearly as long as the four next together ; prothorax rather broadly elongate, covered with numerous large shallow punctures, the sides denticulate, the anterior angle occupied by a strong triangular tooth ; scutellum transversely pentagonal ; elytra closely punctate-striate, becoming gradually darker towards the sides ; legs pale ferruginous ; body beneath dull ferruginous, closely punctured. Length 4 lines.

Brontes nigricans.

B. fuscus ; prothorace lateribus denticulatis, dente antico incrassato ; elytris striato-punctatis, nigricantibus.
Hab. Queensland (Moreton Bay).

Dark ferruginous brown, covered with short, black, setulose hairs ; head slightly exserted ; eyes dark brown : antennæ longer than the body, the first joint shorter than the four next together ; prothorax broadly elongate, rugose, slightly punctated, the sides equally denticulate, the anterior angle occupied by a moderately thickened tooth ; scutellum transversely pentagonal ; elytra punctate-striate, of a uniform dark brown ; legs ferruginous ; body beneath dull ferruginous, closely punctured. Length 4 lines.

From *Brontes denticulatus*, F. Smith (also from Australia), the two species described above differ in the comparatively elongate, not transverse, prothorax and other characters. *Brontes militaris*, Er., is smaller and less robust, narrower prothorax, differently coloured, differently punctured, &c.

Ino [Cucujidæ].

Laporte de Castelnau, Etud. Entom. p. 135.

Ino ephippiata.   (Pl. XVI. fig. 9.)

I. nigra, nitida; elytris disco pallide flavescente, abdominis segmenta tria ultima haud obtegentibus.

Hab. Dorey (New Guinea).

Deep glossy black; head and prothorax about equal in breadth, finely punctured, the latter very much contracted at the base; antennæ half as long as the body, black, the basal joints paler; palpi pale brown; scutellum black, transversely ovate; elytra narrowed at the base, gradually widening posteriorly, where they are as broad as long, the sides straight, the disk with a large pale-yellow spot occupying nearly the whole of the base, except the shoulder, and expanding below the middle towards the side; part of the third and fourth and fifth abdominal segments dull black, not covered by the elytra; legs light glossy-brown, tarsi testaceous; body beneath paler.   Length 1½ line.

Ino trepida.

I. fusca, nitida; elytris singulis flavescente unimaculatis, abdominis segmenta quatuor ultima haud obtegentibus.

Hab. Dorey (New Guinea).

Dark olivaceous brown, shining; head and prothorax equal in breadth, finely punctured; antennæ about one-third the length of the body, the two basal joints yellow, the remainder black; scutellum and elytra as in the last, but the yellow spot on the latter is smaller, nearly round, and situated below the middle and towards the outer margin; abdomen dark brown, shining, the last four segments not covered by the elytra; legs olivaceous brown, the tarsi paler, inclining to testaceous. Length 1½ line.

Ino is a very singular genus, and was placed by M. de Castelnau among the Staphylinidæ, after *Anthobium*. The species described by him (*I. picta*) from Madagascar has slightly elevated lines on the elytra, and it is possible that the two described above may hereafter form another genus.

Phenace [Dasytidæ].

Head short, rounded in front, the epistome and lip concealed beneath its margin. Eyes large, prominent, entire. Antennæ filiform, distant, arising below the eyes, the first joint rather short, obconic, the second very short, the remainder to the tenth longer and subequal, the last longest of all.   Maxillary palpi long, the terminal joint fusiform. Mandibles long, slender. Prothorax rounded at the sides. Elytra broader than the prothorax, elongate. Legs slender; tibiæ spurred; tarsi very long, the basal joint longer than the second.

In general appearance this genus has a wonderfully striking resemblance to some of the Œdemeridæ; its very distinctly five-jointed tarsi, however, independently of other characters, show at once that it can have nothing to do with that family. But there can be no hesitation, I think, in **referring it to** the Dasytidæ, notwithstanding the structure of the mouth **and the** presence of two well-marked spurs to **the tibiæ: in** regard to **the** first, the lip and epistome are so completely hidden by the **scarcely** prolonged anterior margin **of the head,** that, **without** dissection, their **existence can** only be assumed; between this margin and the **mandibles there** intervenes a sort of cavity, **and** the latter, not being covered in the usual way by the lip, are fully exposed almost to their base. **My** specimen, which is unfortunately, I believe, unique, was taken by the well-known traveller Anderson, in **Southern Africa, in the country near Lake** N'Gami.

Phenace œdemerina. (Pl. XVI. fig. 6.)

P. gracilis, fuscescens, parce pilosa; scutello elytrisque pallidioribus.
Hab. N'Gami.

Slender, **dark olivaceous** brown, sparsely **clothed** with **rather long,** pale-greyish hairs; head and prothorax shining, dark brown; scutellum elongate, rounded below, a depressed longitudinal line in the middle; elytra narrow, elongate, nearly parallel, the shoulders rather prominent, **substriate, olive-brown, paler as it** recedes from the base; mandibles bright ferruginous; **legs reddish brown; body beneath** dark brown, hairy. **Length** 3¼ lines.

Оснотуга [Lampyridæ].

Head partially exposed, short, broad in front. Eyes very large, contiguous beneath, **constricted behind.** Antennæ very short, 12-jointed, the two basal thickened, the rest serrated. Prothorax transverse, narrower than the head. Elytra broader than the prothorax, subparallel, shorter than the abdomen. Legs moderately short, all the coxæ nearly contiguous; tarsi slender. Abdomen eight-jointed in the male, the joints gradually decreasing in breadth to the apex.

This **genus is allied to** *Dioptoma* (ante, p. 118), **and the** nearest affinity **of the two is** apparently with *Luciola*, Lap. (*Colophotia*, Dej.). **In** the only example I **have seen of the former** the abdomen has been removed, **but, judging** it from what we now see of this, it is probably also exserted, with the same number of segments—the normal number, in fact, **in the males. The** females of both are unknown.

Ochotyra seminusta. (Pl. XVI. fig. 7.)

O. pallide fulva; capite prothoraceque piceo-fuscescentibus.
Hab. India (Malabar).

Pale fulvous yellow, **very** sparsely covered with greyish appressed hairs; head pitchy-brown, concave between the eyes, epistome with

stiff greyish hairs; antennæ not extending beyond the eyes, pale
yellow, strongly serrated, broadest in the middle; eyes dark brown,
shining; prothorax light pitchy brown, darker on the disk; scutellum
rather large, triangular; elytra about twice the length of the head and
body together, depressed, and almost concave posteriorly, with elevated
nervures in the middle, *i. e.* not extending to the base or apox; legs
clothed with stiff hairs, particularly on the tibiæ; abdomen dull
whitish yellow. Length 4 lines.

In the Plate the figure of this species is longer than it ought to be.

Ethas [Tenebrionidæ].

Head elongate, broader than the prothorax, rounded and dilated anteriorly,
narrowed into a neck behind. Eyes remote from the prothorax, lateral,
partially divided posteriorly. Mentum somewhat pentagonal, narrow
at the base, concealing the labium. Maxillary palpi robust, the terminal joint subcylindric, of the labial ovular. Antennæ stout, eleven-
jointed, the first largest, the second shorter than the third, which, with
the remainder to the tenth inclusive, are transverse and cup-shaped, the
eleventh small, shortly cylindric. Prothorax sulcated, subquadrangular,
broadest in front, the anterior angles rounded, the sides keeled. Elytra
elongate-ovate, wider than the prothorax, ribbed. Legs robust; femora
slightly clavate; tibiæ not spurred; tarsi ciliated beneath; prosternum
produced, rounded anteriorly.

The ribbed prothorax and elytra will at once distinguish this genus
from *Stenosis*, which has exactly the same habit. The structure of
the mouth varies a little from that genus, in *Ethas* the large angular
mentum filling up more of the oral cavity, and entirely concealing
the labium.

Ethas carbonarius.   (Pl. XVI. fig. 2.)

E. niger, subnitidus; prothorace leviter trisulcato; elytris singulis lineis
quinque elevatis instructis.
Hab. Malabar.

Punctured, black, slightly shining; head convex between the eyes,
with three rather shallow grooves, and on each side a somewhat deeper
groove in which the eye is placed; prothorax a little narrower than the
head, marked with three lightly impressed lines or grooves, the spaces
between, especially the two middle, slightly elevated and convex; scutellum punctiform; elytra scarcely wider than the prothorax at its
base, each with five narrow elevated lines rather thickened at the
suture, but scarcely forming another; legs slightly pitchy; palpi ferruginous; body beneath black, shining, sparingly punctured. Length 4
lines.

Ethas stenosides.

E. niger, subnitidus; prothorace profunde trisulcato; elytris singulis
lineis quatuor elevatis instructis.
Hab. Siam.

Punctured, black, slightly shining; head very convex between the eyes, not sulcated; prothorax much narrower than the head, only slightly dilated anteriorly, with three broad and deep longitudinal grooves, the spaces between sharply elevated (with the keeled sides forming altogether four narrow but prominent costæ); scutellum small, triangular; elytra broader than the prothorax at the base, each with four raised lines, the two central abbreviated towards the apex, the suture not thickened; legs pitchy; body beneath black, sparingly punctured; antennæ, especially towards the apex, sparsely clothed with rich golden-brown hairs. Length 2⅔ lines.

Smaller and proportionably narrower than the last, and readily distinguished by a multitude of characters, although the habit is nearly the same.

Aposyla [Tenebrionidæ.]

Head convex and subtriangular in front, slightly elongated behind the eyes. Antennæ short, eleven-jointed, gradually increasing from the base, the first joint partially concealed by the antennary orbit. Eyes large, round, entire. Epistome and lip short, very transverse. Palpi with the terminal joint narrowly triangular. Mentum transverse. Prothorax subcordate, scarcely longer than broad. Elytra narrow, sub-parallel. Legs moderate; anterior coxæ large, subcylindrical, greatly exserted; tibiæ spined; tarsi slender.

But for the large and greatly exserted anterior coxæ, I should not hesitate to place this genus near *Calcar*, although the antennary orbit is so contracted as to leave the eye perfectly free, and the epistome, although short, is of great breadth and apparently distinct from the front. Whatever its affinities may be, I cannot myself see, at present, that it can be better placed than near *Calcar* and *Boros*.

Aposyla picea. (Pl. XVI. fig. 4.)

A. subelongata, rufo-fusca, nitida, punctata; antennis ferrugineis.
Hab. Queensland.

Rather elongate, subdepressed, shining, reddish brown; head convex between the eyes, and slightly constricted behind them, irregularly punctured; antennæ ferruginous; lip with stiff greyish hairs; prothorax with numerous somewhat coarse punctures; scutellum broadly triangular; elytra scarcely wider than the prothorax, punctured in rather irregular rows; body beneath and legs reddish-brown. Length 3½ lines.

Rhypasma [Tenebrionidæ].

Head rather broad, convex in front, truncate anteriorly, the epistome and lip inserted beneath. Eyes small, oblong, entire. Antennæ 11-jointed, inserted beneath the broad antennary orbit, half the length of the body, the first three joints longer, the next five submoniliform, the last three forming a narrow club. Mentum large, transverse. Prothorax longer than broad, subquadrangular, narrowed behind, sinuated in front, longitudinally sulcated. Elytra subdepressed, carinated, scarcely broader than the prothorax, and slightly rounded at the side. Legs rough, moderately robust; tibiæ fusiform; tarsi narrow; the claw-joint as long as the rest together; pro- and mesosterna simple; post-intercoxal plate broadly truncate anteriorly.

Notwithstanding the small size of this insect compared with *Zopherus* and *Nosoderma*, there can be little hesitation, I think, in placing it near those anomalous genera. Judging from the examination of the oral organs made *in situ*, they appear to offer only a slight modification of those of *Nosoderma*, the mentum, however, being considerably larger and in great measure hiding the palpi and base of the maxillæ, the part between its lateral margin and the insertion of the antenna offering a deep cavity, as in that genus, for the reception of its basal joints when that organ is in repose. The propectus has no antennary canal at its side as in *Zopherus*, in this respect agreeing better with *Nosoderma*; on the other hand, the latter has only a ten-jointed antenna, but this is again modified by the fact that *N. obcordatum*, Kirby, has eleven.

Rhypasma pusillum. (Pl. XVI. fig. 3.)

R. obscure testaceo-brunneum; prothorace trisulcato; elytris disco tricostatis, costa intermedia abbreviata.
Hab. Para.

Dull testaceous brown, more or less sprinkled with a semicrystalline exudation? head with numerous small granules, and having the appearance of being originally covered with an earthy crust; antennæ covered with granulations, each tipped with a fine hair; prothorax with two curved longitudinal costæ on the disk, nearly meeting anteriorly, the lateral margins flattened and resembling the costæ, and like them crested with a number of small closely set granules, the spaces between the costæ and the margins respectively forming three broad shallow grooves; scutellum transverse, subquadrate; elytra rounded at the shoulder and at the apex, the disk with three strongly crenulated costæ, the outer and inner united near the apex, the intermediate ceasing at two-thirds the length of the others, the external margin also forming a crenulated border, resembling the costæ, the spaces between deeply and coarsely punctured; legs covered with small asperities and cilia; body beneath rufous-brown, covered with numerous granulations. Length 2 lines.

With reference to what is probably an exudation (renewable perhaps at the pleasure of the animal), it has the appearance under the microscope of small particles of brown sugar.

CHARTOPTERYX [Helopidæ].

Westwood, Arc. Entom. i. p. 43.

Chartopteryx binodosus.

C. obovatus, **fusco-cupreus**; elytris basi bigibberis.
Hab. Queensland.

Obovate dark copper-brown, irregularly punctured with **numerous** nearly erect hairs arising from the punctures; head with a **transverse** impression above the epistome, roughly but rather sparingly punctured; antennæ black, not reaching beyond the base of the prothorax, the four terminal joints dilated; prothorax transverse, sinuate in front, anterior angles produced, **the posterior rather acute**, with shallow scattered punctures; scutellum subtriangular; **elytra very** convex, a large compressed elevated protuberance near the base of each, rather dilated, **posteriorly covered with large rough punctures**; legs hairy; **body beneath less coppery and more slightly punctured**, with fewer hairs. Length 5 lines.

This species differs considerably in habit from *C. Childrenii*, West., and in that respect bears a marked resemblance to *Thecacerus binodosus*, Lap., belonging to the same family. It may be necessary eventually to propose a new genus for its reception.

CYPHALEUS [Helopidæ].

Westwood, Arc. Entom. i. p. 43.

Cyphaleus insignitus.

C. ovatus, niger, **subnitidus**; **elytris viridi-metallicis**, nitidissimis.
Hab. Queensland.

Ovate, everywhere black except the elytra, slightly shining on the prothorax and beneath, the upper surface irregularly covered with deep round punctures, most numerous on the sides of the prothorax posteriorly and **base** of the elytra, but which gradually disappear towards the apex, the punctures small on the head, prothorax, and scutellum, but nearly all with a stiff setose hair arising **from** the interior of each; elytra very convex, dark metallic green with purple and violet reflections; legs bluish black, finely punctured. Length 9 lines.

Probably most akin to *C iopterus*, **Westw.**, but is narrower, with **the prothorax more** convex, &c. In *C. iopterus*, too, the prothorax **is a dark** metallic green, and the elytra a very deep purple with violet reflections, particularly at the sides. The figure of Professor Westwood in the 'Arcana,' at pl. 12. f. 1 (not referred to in the text), somewhat resembles the present, but is certainly not applicable to either of the three species there enumerated.

OSDARA [Helopidæ].

Walker in Annals and Mag. Nat. Hist., 3 ser. ii. p. 284.

Osdara lævicollis.

O. capite prothoraceque nigris, lævibus; elytris subferrugineis, nigro
tuberculatis; pedibus rufo-ferrugineis.

Hab. Ceylon.

Ovate, convex; head and prothorax smooth, glossy black, very finely
punctured; scutellum small, triangular; elytra pale ferruginous with a
tinge of grey, covered with irregular lines of black tubercles which,
under the lens, are seen to be composed of smaller ones (from 2–10);
amongst these, in the intervals, a few deeply impressed punctures; legs
bright reddish-ferruginous, anterior and intermediate tibiæ with a small
rounded tooth near the extremity internally; antennæ at the base and
palpi ferruginous; body beneath dark brown, coarsely punctured. Length
4 lines.

This very interesting and distinct species agrees generically, except as regards the mouth, which has not been examined, with *O.
picipes*, save in the toothed tibiæ, which in this instance can only be
considered of secondary importance. Both species have more or
less of a gloss, which has the appearance of being due to varnish:
the black shining prothorax of the present, however, contrasted
with the elytra, is very marked, and recalls many *Adesmiæ*, to which
also it is very similar in form. A single specimen sent by Mr.
Thwaites from Ceylon is in my collection.

OZOTYPUS [Helopidæ].

Characters nearly as in *Osdara*, Walker*, but differs in the epistome not being separated from the front by any groove, by the
absence of the scutellum, by the form of the tibiæ, which are fusiform and attenuated most at the extremity, and by the shortness of
the tarsi, the claw-joint being as long as the rest together. As
secondary characters, the form is narrower and more convex, the
antennæ shorter, and the prothorax gibbous anteriorly. As in
Osdara, the prosternum has a sharp-keeled process which is received
into a corresponding notch of the mesosternum, and the intercoxal
plate is broad and rounded anteriorly. The same varnished appearance is also as noticeable, but only on the elytra. In *Ozotypus* the
tubercles which cover the upper surface are smaller, more regularly
arranged, and each tipped with a short curved hair, which is not the
case in *Osdara*. In both genera the tarsi are all nearly of equal

* For a more detailed description of *Osdara*, see Lacordaire, Gen. de Coléopt.
v. p. 455.

length, and the penultimate joint is shorter and narrower than the
preceding ones.

Ozotypus setosus.

O. ferrugineus, tuberculatus, tuberculis setigeris.
Hab. Ceylon.

Subovate, ferruginous, almost everywhere covered **with** setigerous
tubercles, except the epistome and antennæ; head rather small, slightly
concave in front; prothorax transverse, rounded at the sides, produced
into an angle anteriorly, a prominent gibbosity in front partially **over-**
hanging **the** head, and irregularly studded with granular tubercles;
elytra nearly ovate, wider than the **prothorax** at the base, the tubercles
closely and regularly arranged **in lines (nine or** ten **on each);** legs
reddish-ferruginous, rather short, **slender, closely covered** with small
tubercles bearing rather longish setæ; **tarsi very short, the** basal joint
shortly triangular, the rest, except the **last, very transverse** and clothed
with sparse stiff hairs; antennæ **rather more than a fourth as long as**
the body, **more** claviform **than in** *Osdara picipes;* eyes brown; mandibles **dusky;** body beneath dull ferruginous. **Length 4 lines.**

Apolecta [Anthribidæ].
Pascoe, Ann. and Mag. Nat. Hist. 3 ser. iv. p. 431.

Apolecta fucata.

A. pallide grisea, nigro varia; capite prothoraceque griseo bivittatis;
elytris maculis approximatis; tarsorum articulo primo basi cinerascente.
Hab. Ceram.

Narrowly oblong, with a short pale-greyish pile varied with black;
head and prothorax black, with two greyish or dull-white stripes from
between the antennæ and eyes, and terminating at the posterior border
of the latter; antennæ three to four times as long as the body, black,
the last **three joints white;** eyes dark horn-colour; prothorax longer
than broad, narrowed in front; scutellum small, transversely oblong;
elytra subovate, dull greyish, with large black approximate or confluent
patches; legs black, the first joint of all the tarsi ashy above at the
base; body beneath dark brown, slightly shining, margin of the metasternum and of all the abdominal segments greyish. Length 7–8 lines.

This is the largest and most robust of all the described species,
and nearest in colour to *A. parvula*, Thoms. The spots on the
elytra are more or less confluent according to the individual.

Mecocerus [Anthribidæ].
Schönherr, Gen. et Sp. Curcul. i. p. 115.

Mecocerus insignis.

M. robustus, griseo-ochraceus, atro maculatus; prothorace paullo longiore
quam latiore; antennis pedibusque atris.
Hab. Ceram.

Robust, with a pale-greyish ochraceous pile spotted with black; head
with two hairy ochraceous stripes in front, the sides below with deep,
coarse punctures; prothorax a little longer than broad, slightly narrowed
behind, ochraceous, with more or less confluent black spots; scutellum
obscure ochraceous; elytra subparallel, convex, ochraceous, with small
black spots, seriate-punctate; legs black, robust, and elongate in ♂,
with the basal anterior tarsal joint longer than the succeeding ones
(of equal length in ♀, with the two intermediate joints not longer than
the claw-joint); body beneath black, the sides of the metasternum and
abdomen with a double row of dull ochraceous spots; antennæ black,
robust, and three times as long as the body in ♂ (not reaching to the
base of the prothorax in ♀). Length 12 lines.

M. variegatus, Ol., is distinguished from this by its pale-ashy
pubescence, narrower form, antennæ scarcely twice the length of
the body, and larger spots. It is not impossible, however, that this
may turn out to be only a strongly marked local sub-species.

Mecocerus maculosus.

M. subelongatus, griseo-ochraceus, atro maculatus: prothorace longiore
quam latiore, postice attenuato; antennis pedibusque atris.
Hab. Ceram.

Rather elongate, pale greyish, slightly tinted with ochraceous, and
spotted with black; head with two hairy, greyish stripes between the
eyes, the sides below obscurely punctured; prothorax much longer
than broad, narrowed behind, black, a central stripe and two spots on
each side greyish; scutellum black; elytra subparallel, slightly depressed, ochraceous with large black spots, seriate-punctate; legs black,
in ♂, moderately elongate, slender, the first anterior tarsal joint not
longer than the succeeding ones together (in ♀ the two intermediate
tarsal joints longer than the claw-joint); body beneath black, the sides
of the metasternum and abdomen with a double row of dull ochraceous
spots; antennæ black, robust, nearly three times as long as the body
(in ♀ extending beyond the base of the prothorax). Length 8 lines.

On a superficial examination this might be taken for a small
variety of the above; but in addition to the distinctions noted in the
description, it may also be observed that the spots in this species are
much larger and form a less numerous series along the suture. Mere
colour, in the Anthribidæ, is not to be depended on unless accompanied by a certain variation of pattern, as is the case in this
instance.

Mecoerus allectus.

M. subbrevis, griseo-fulvus nigroque varius; antennis pedibusque nigris,
his griseo annulatis.
Hab. Cambodia.

Rather short, with a greyish-yellow pile; head black, an elongato-

obcordate yellowish spot on the vertex, descending between the eyes
but not passing beyond them ; antennæ nearly three times as long as
the body in ♂, black, the intermediate joints greyish at the apex ;
prothorax as long as broad, an irregular patch on the disk, apparently
made up of smaller spots, and occupying its whole length, occasionally
two or three smaller spots at the side ; scutellum small, triangular,
black ; elytra short, broadly ovate, greyish yellow, with four principal
spots on the disk, the shoulder, a few smaller spots at the sides (sometimes nearly obsolete), and several at the apex black ; legs of moderate
length, the anterior but little produced in ♂, femora and tibiæ obscurely
ringed with grey, the basal and claw-joints of the tarsi ashy, except at
the apex ; side of the propectus, metasternum, and abdomen closely
covered with a yellow pile, a spot on each side of the segments and
the middle of the apical one black.   **Length 9 lines.**

In none of the species described above have the males a spined
propectus.

Dœothena [Anthribidæ].

Head small, not contracted below the eyes, rostrum very short. Antennæ 12-jointed, very slender, much longer than the body, arising
from a cavity beneath and a little in front of the eye, the first joint
swollen at the base, gradually diminishing upwards, and terminated in
a truncated apex, the second as long as the first, but slenderer and
obconic, the remainder to the eighth inclusive subequal, filiform, their
apices more or less tumid, the apical third of the ninth and three terminal joints forming an oblong slender club. Eyes large, lateral, deeply
emarginate beneath. Antennary cavity grooved above. Epistome and
lip forming together a small triangle covering the centre of the mandibles. Palpi filiform. Prothorax convex, rounded anteriorly, as wide
as the elytra at the base, the carina immediately in contact with the
base at the middle, but slightly and gradually diverging towards the
side, forming a sharp angle at its flexure, then continued to half the
length of the prothorax, where it suddenly ceases. Elytra convex, not
gibbous at the base. Legs of moderate length, first tarsal joint elongate.

The insect which has served for the above generic description is
exceedingly like *Protædus mærens*, Pasc.[*] On examination, however, they will be found to be not even generically identical, the
twelve-jointed antennæ and its club, composed not of three only but
also by part of a fourth joint, being, I believe, unparalleled among the
Anthribidæ, and the emarginate eye and the position of the abbreviated carina being quite different in *Protædus*. I do not here more
than allude to the enormous size of the two intermediate tarsal joints,
as it is just possible that that may be only a sexual character. The

* Ann. and Mag. Nat. Hist. 3rd series, v. p. 39.

emargination of the eye corresponds to a kind of groove in the upper portion of the antennary cavity, and is obviously intended to allow the antennæ to be thrown well back; this structure does not exist in *Protœdus*. Another peculiarity is the form of the basal antennary joint, which has a pyriform shape, but with the small end at the apex, which is the reverse of what generally occurs; but some slight approach to this is made in *Protœdus*, where the greatest diameter is in the middle.

Dœothena platypoda.   (Pl. XVI. fig. 1.)

D. elongato-ovata, nigro-pubescens, albo varia; tarsorum articulis duobus intermediis peramplis.
Hab. New Guinea (Mysol).

Elongate-ovate, somewhat sparsely covered with dull black, varied with white, coarsish, slightly curved hairs; head with the pubescence nearly entirely white; prothorax with two large black patches on the disk, divided by a very narrow median line, and two smaller ones on each side; scutellum rounded below, very indistinct; elytra obsoletely punctate-striate, the sides and middle black, the basal and apical portions white with a few oblong black spots; pygidium white; antennæ dark brown, paler at the base; eyes and mandibles black; maxillæ, palpi, and labrum rufous; body beneath and legs white. Length 2½ lines.

Owing to the somewhat sparse pubescence, the darker ground is seen beneath the white hairs, thus giving them a pale-ashy hue. The appearance of the markings seems to show that the proportion of the two colours may vary.

Picænia [Anthribidæ].

Head rather broad in front, the rostrum very short, slightly emarginate at the apex for the insertion of the small epistome and lip. Antennæ short, eleven-jointed, arising from a cavity beneath the rostrum and close to the eye, the first two joints ovate, thickened, the remainder to the eighth inclusive more or less conic, the last three forming an ovate, compact, depressed club. Eyes large, round, nearly entire. Palpi slender, hairy, the last joint of the maxillary fusiform. Prothorax transverse, rounded in front and at the sides, the carina basal, and terminating close to the anterior border of the prothorax. Elytra short, convex, parallel to the base of the prothorax. Pygidium small, narrow. Legs rather short. Tarsi short, the basal joint scarcely larger than the intermediate two. Claws strongly toothed at the base.

A short convex form, very much resembling *Misthosima* in appearance, but differing in the subrostral insertion of the antennæ, the ovate compact club, short tarsi, and other characters. The short, or rather, perhaps, the entire absence of rostrum will at once prevent its being confounded with any genus having its attenuated club of the same form, such as *Ethneca*, *Penestica*, or the females of *Anthribus*.

Piarnia saginata. (Pl. XVI. fig. 8.)

P. breviter ovata, pube nigra albo maculata vestita; elytris vage seriatim punctatis.

Hab. Borneo.

Shortly ovate, covered with a close black pile with white spots; head nearly circular in front, but a little narrowed below the eyes, no raised line, a few white hairs mixed with the black; antennæ not longer than the breadth of the head, black, the club occupying rather more than a third of the total length; prothorax as broad as the elytra, black, passing into white at the sides, with a few white spots on the disk; scutellum transverse, white; elytra black, irregularly spotted with white, especially near the base and apex; body beneath and legs with a close greyish-white pile. **Length 2 lines.**

ZYGÆNODES [Anthribidæ].

Pascoe, Ann. and Mag. Nat. Hist. 3 ser. iv. p. 328.

Zygænodes monstrosus. (Pl. XVI. fig. 5.)

Z. fuscus, sparso griseo pubescens; elytris singulis fasciculis tribus prope suturam sitis.

Hab. Natal.

Dark brown, with a sparse greyish pile obscurely clouded with dull fulvous; head a little broader than the prothorax, flat and triangular in front, uniformly of an obscure grey; prothorax nearly twice as broad as long, the disk irregular, subquadrituberculate; the carina prominent; scutellum triangular, pale grey; elytra not broader than the prothorax, irregular, punctate-striate, on the disk a few raised points, which are rather darker than the rest, and on a line parallel to the suture three dense fascicles of pale-greyish hairs, the first and largest near the base, the other two towards the apex; body beneath brown, with greyish hairs; legs dull testaceous, with darker rings; antennæ pale greyish yellow, the third joint, upper part of the fourth, and fifth near the apex, and the last three forming the club, black; eyes dark brown. Length 1½ line.

This curious Anthribid, agreeing generically with *Zygænodes*, differs remarkably in colour as well as in the irregularity of its surface from *Z. Wollastoni*; but that a genus so peculiar should be represented in countries so far apart, although by no means singular, is a fact well worthy of note. It is probable that hereafter the genus may be found to be rich in species; there are two new ones in Mr. Bowring's extensive Asiatic collections, as well as numerous others belonging to genera which I have proposed in this Journal and elsewhere, and which now contain each but a single represen-tative.

NESSIARA.

Pascoe, *ante*, p. 60.

Nessiara scelesta.

N. fusca, pubescens; prothorace elytrisque planatis, his singulis bituber-culatis, macula magna communi nigra.

Hab. Island of Mysol (New Guinea).

　　Clothed with very short, dark-tawny-brown hairs; head finely punc-tured, a single short central carina on the rostrum, dark brown passing into black at the mouth and mandibles; antennæ not longer than the rostrum, brown, the two basal joints yellow; prothorax flattened above, the depressed portion at its junction with the side forming a sharp, irregular, dark-brown or black line; scutellum small, transverse; elytra short, seriate-punctate, the disk depressed, somewhat concave, having an obtuse spreading tubercle at each angle, the middle of the depres-sion with a large subquadrate black patch; legs dark brown, the tibiæ and tarsi ringed with grey; body beneath brownish black, with a very thin greyish pubescence. Length 4 lines.

Not quite so much depressed as *N. planata* (*ante* p. 60), with the median patch of *Nessa centralis* (Ann. and Mag. Nat. Hist. 3 ser. iv. p. 329), and much darker than either of them. Among the undescribed species in Mr. Bowring's collections, there is one with the sides of the rostrum dilated in a most extraordinary manner.

GOËPHANES [Lamiidæ].

Head quadrate in front. Eyes small, lateral, reniform. Antennæ se-taceous, longer than the body, arising from short, moderately distant tubercles, the basal joint rather elongate, subcylindrical, the third long-est, the remainder gradually shorter. Epistome and lip very short, transverse. Mandibles entire at the apex. External maxillary lobe elongate. Prothorax subovate, unarmed. Elytra rather depressed, broadest at the base, the sides rounded, the apex oblique. Legs mode-rate, femora clavate, tarsi slender, the basal joint of the four posterior elongate. Pro- and mesosterna simple.

In habit this insect resembles *Glaucytes*, but is a true Lamiid, although its exact affinity is not very obvious; for the present, how-ever, I am disposed to place it among the Acanthocinæ, perhaps near *Liopus* or *Œdopeza*.

Goëphanes luctuosus. (Pl. XVII. fig. 2.)

G. ater, albo variegatus; antennis atris; articulis quarto, apice excepta, et ultimis quatuor albis.

Hab. Madagascar.

Pubescent, deep black, varied with a nearly pure white (the figure
will give a better idea than any description), a few bristly hairs fring-
ing the sides of the elytra; tarsi brownish, and the terminal joint
yellowish white; antennæ slightly ciliated beneath, black, the fourth
joint, except at the apex, and last four joints white; body beneath
pitchy, with a sparse silvery pile. Length 4 lines.

AGELASTA [Lamiidæ].

Newman, Entomologist, p. 288.

Agelasta Mouhotii.

A. cinereo fulvoque pubescens, **nigro maculata**; elytris fasciis duabus
fuscis ornatis; tibiarum apice tarsisque nigris.
Hab. Cambodia.

Sparingly pubescent, **the** dark shining epiderm everywhere more or
less visible; head and prothorax with a thin fulvous pile, spotted with
**dark brown on the latter; the fulvous passes into ashy posteriorly, and
is continued on to the scutellum and base of the elytra, where it is
limited by a broad band of dark brown, having its posterior border**
very irregular; the rest of the elytra is fulvous with a denticulate band
towards the apex; **and** the whole, not occupied by the two bands, is
dotted with small **brown or** nearly black spots, the centres of each
being occupied by a shallow **puncture**; legs ashy, the lower half of the
tibiæ and the tarsi black; antennæ scarcely longer than the body,
black, the first three, base of the fourth, and the fifth joints ashy; body
beneath with a thin ashy **pile.** Length 5-6 lines.

This very distinct species, which is perhaps most nearly connected
with *A. amicus,* Wh., **may be recognized by the clear** ashy-grey **at
the base of the elytra, contrasted with** the **rich-dark-brown band
which succeeds.** I have dedicated it **to M.** Mouhot, who, **as is well**
known is now, and has been for some years, **investigating the Zoo-
logy of Cambodia and Siam.**

Agelasta rupta.

A. obscure-griseo pubescens, **nigro maculata**; elytris fasciis duabus dentatis nigris; tibiis annulatis tarsisque nigris.
Hab. Cambodia.

Sparingly pubescent, **dull** greyish, spotted and banded with black;
head yellowish grey, obscurely spotted; antennæ longer than the body,
the first two and basal half of **the third** joint grey, the remainder
black, with the fourth, sixth, eight, tenth, and eleventh at their bases
more or less ashy; prothorax very short and transverse, yellowish grey,
spotted with black; scutellum nearly quadrate, the **apex** slightly
rounded; elytra short, subparallel, irregularly punctured, greyish, a
toothed band between the base and middle, and a narrower waved in-

terrupted one towards the apex, with several spots, black, each band bordered with dull fulvous; femora and tibiæ greyish, ringed with black, tarsi black, base of the claw-joint only grey; body beneath dull brown, with a very thin greyish pile. Length 5 lines.

Resembles the last in colour, only it is much less pure, and the elytra has not the ashy base of that species. The prothorax is unusually short for an *Agelasta*.

Agelasta catenata.

A. piceo-fusca, pilosa, atra, murino alboque lineata; antennis pedibusque annulatis, illarum articulis terminalibus brevibus, ciliatis.

Hab. Cambodia.

Pitchy brown verging to black, with a short close pile, running in narrow, longitudinal, irregular and partially interrupted lines of brownish grey and white, bearing similar lines, or here and there on the elytra spots, of the black epiderm between them; antennæ scarcely longer than the body, more or less brown and black, the fourth to the seventh joints inclusive white at the base, the apex of the latter and the remainder (which are much shorter) densely ciliated beneath; legs greyish white varied with brown; tarsi greyish white, the apex of the fourth joint and claws black; body beneath pitchy black, with a greyish-white pubescence. Length 7½ lines.

The specimen described above is probably, from the structure of the antennæ, a female; the same crowding together of the terminal joints is seen also, and in the same sex, in *A. polynesus*, White. Like the last, it was sent from Cambodia by M. Mouhot.

Niphona [Lamiidæ].
Mulsant, Longic. de France, p. 169.

Niphona suffusa.

N. fusca, undique pubescens, supra variegata; prothorace irregulari, lateribus tuberculis duobus distantibus; elytris basi tuberculo parvo instructis, humeris elevatis.

Hab. Cambodia.

Robust, dark brown, covered with short closely set hairs; head slightly gibbous between the eyes, with an impressed longitudinal line, rusty yellow, more or less varied with dark brown; prothorax transverse, narrow anteriorly, the disk irregular, bituberculate at the side; an impressed line posteriorly, rusty yellow, with three longitudinal bands on the disk; scutellum transverse, rounded below, black, the sides paler; elytra much broader than the prothorax, gradually narrowing from the shoulders, which are very prominent and produced anteriorly, a small tubercle at the base, covered with pale-yellowish hairs, and irregularly spotted with black, particularly at the base, where they

become more or less confluent, more crowded also towards the apex
and at the sides, bordered, particularly on the suture, by a rose-red
line, three longitudinal lines of the same colour on each, the middle **and**
exterior extending nearly to the apex; legs rose-red, annulated with
black, the last two tarsal joints black; body beneath rose-red, the
centre of each abdominal segment black at the base; antennæ with
the basal joint shorter than the third, black, the two first and base of
the remainder rose-red. **L**ength 11 lines.

Rather larger than *N. thoracica*, Wh., to which it bears a general
resemblance, but distinguished by the comparative regularity of the
disk, and the absence of the peculiar medio-basal fissure of the prothorax.

Niphona pannosa.

N. subangustata, grisescente **tomentosa, variegata**; prothoracis lateribus
tuberculiferis; elytris basi piloso-cristatis; **tibiis anticis rectis.**
Hab. Cambodia.

Rather narrow, **the male** broader, **covered with a dense, very pale-**
greyish tomentum, spotted with a darker or mouse-coloured **grey**;
head rather small; eyes and lip black; antennæ rather more than two-
thirds the length of the body, dark grey with very pale spots; prothorax narrower than the elytra, subtransverse, irregularly tuberculate,
especially towards the base, the side with a few short tubercles, partially disposed in two **rows**; scutellum transverse; elytra broadest at the
shoulders, gradually narrowing towards the apex, irregularly costulate
especially towards the apex, coarsely punctured, the base on each side
with **a short**, narrow, erect **tuft of** hair, the apex subtruncate, pale
greyish, darker posteriorly, **so as** to appear as a band, the shoulder
sometimes dark brown; legs closely covered with short hairs, pale,
spotted with darker grey; abdomen hairy **at** the sides, with dark-grey
spots, the sterna reddish **brown with** paler spots. **L**ength 8 lines.

N. cylindrica, **White, differs in** its extraordinary fore tibiæ, and
in its greatly developed lateral tubercle; **and** *N. Ferdinandi*, Paiva,
in the absence of the basal crest of the **elytra, &c.** In the latter
species the claw-joint is scarcely **half** the length **of** the three preceding, while in others it is as long as the rest together,—another
instance of the shifting characters **of** the Longicorns, and so far of
greater importance as **the** large claw-joint generally marks its possessor to be a "twig-climber," in distinction to the short-clawed
species, which are principally found on the *trunks* of trees.

Niphona excisa.

N. angustata, nigra, pube grisea tecta; prothorace profunde trisulcato;
elytris postice attenuatis, apice divaricatis, singulis fortiter emarginatis
Hab. Cambodia.

Narrow, subcylindrical, black, covered with a short, thin, greyish pile; head rather short, narrowed below the eyes, the vertex lengthened; lip and epistome small; mandibles black, palpi ferruginous; eyes (for *Niphona*) large, black; antennæ shorter than the body; prothorax about equal in length and breadth, constricted anteriorly, the lateral tubercle obtuse, with an indeterminate base; the disk deeply trisulcate, with three or four shorter sulci on each side; scutellum very transverse; elytra coarsely and remotely punctured, a little broader than the prothorax at the base, gradually tapering in nearly a straight line to the apex, which is shortly divaricate and very deeply emarginate, with the two apiculi formed by the emargination nearly equal in size and much produced, the base with two short crests, the inner pilose, at the apical third an oblique indistinct buffish patch; legs and body beneath covered with long greyish hairs. Length 8 lines.

At first sight this species might be readily taken for *N. Ferdinandi*, Paiva; they are, however, abundantly distinct. Touching only a few characters, it may be remarked that the shorter head and larger eye brings this latter organ in pretty close approximation to the base of the mandibles; the palpi ferruginous, not pitchy black; on the prothorax the sulcations are deeper, and the two central elevated lines are entire; the elytra are longer and narrower, the apex shortly divaricate, the emargination very considerably broader and deeper, and the inner as well as the outer apiculus equally prominent and pronounced (in *N. Ferdinandi*, the inner apiculus is sloped away obliquely); there are also the two crests at the base, and the patch posteriorly on the elytra, no trace of either of which exists in *N. Ferdinandi*. Numerous specimens of both species have been received from M. Mouhot.

Niphona arrogans.

N. fusca, griseo pubescente varia; prothorace transverse sexcristato; elytris rude punctatis, basi latis, apice sinuatis.
Hab. Borneo.

Robust, dark brown, with a short, varied, greyish pile; head greyish, with a few scattered punctures; antennæ shorter than the body, brown varied with grey, particularly at the bases of the third and succeeding joints; prothorax transverse, narrower anteriorly, bituberculate at the side, the disk with a series of six short, longitudinal crests, forming a curved line sweeping round from the two lateral tubercles to near the base, greyish, darker or more fulvous posteriorly; scutellum small, very transverse; elytra rugosely subplicate longitudinally, with numerous coarse crowded punctures, broad at the base, tapering gradually behind, the apex sinuate; legs short, varied with grey and brown, the intermediate and posterior tibiæ black at the apex externally, claw-joint as long as the rest together; body beneath with a pale-greyish pile. Length 10 lines.

The crescent-shaped series of short crests on the prothorax will readily distinguish this species; the hairs on the elytra appear to be very deciduous, and are generally rubbed off the more prominent portions.

SYMPHYLETES [Lamiidæ].

Newman, Entomol. p. 362.

Symphyletes pubiventris.

S. subcylindricus, pube cinerascente fulvaque varius; elytrorum lateribus maculis duabus albis; maris abdominis segmento secundo ampliato, densissime hirsuto.

Hab. Australia (Kangaroo Island).

Subcylindrical, black, covered with a short, very pale ashy pile, varied with light fulvous, and spotted with coarse black punctures; head rather narrow, the vertex very convex; antennæ nearly equal in both sexes, not so long as the body, dark brown, not spotted, and very slightly ciliated beneath; prothorax nearly equal in length and breadth, the anterior margin scarcely narrower than the posterior, the side a little rounded, although irregularly, the disk with the two usual shallow transverse depressions; scutellum subtriangular, rounded posteriorly; elytra subparallel, the apex entire, several black shining granules arranged in irregular rows, and extending to near the apex, two white irregular spots on each side partially margined with dark brown; legs and body beneath covered with a similar varied pile; the second abdominal segment in the male larger than in the female, and densely covered with short erect hairs. Length 8 lines.

A more cylindrical species than most others of this genus, in general colour approaching *S. fronticornis,* Fab.; but the two white spots on the sides of the elytra will readily distinguish it. The peculiar structure of the second abdominal segment is very rarely met with among the Longicorns, and appears to be confined to the males. There is nothing to distinguish *Symphyletes* from *Rhytiphora,* Serv., except that the latter has not the lateral tooth on the prothorax, which generally characterizes the former; tho last joint of the antennæ, "*apice repente curvato,*" which Newman gives as a character, is only found in two or three species. How *Penthea,* Lap., is to be distinguished I don't know; it is a stouter form, with shorter and more robust legs, than either *Symphyletes* or *Rhytiphora.* My *Penthea conferta* (Aru), from its toothed mesosternum and absence of antennary tubercles, must be excluded from the genus. Perhaps it should be placed near *Coptops,* Serv. The spine on the anterior coxæ of the males is confined to *S. pedicornis*, Fab., and *S. metutus*, Pasc., and is absent in *S. nodosus*, Newm. (the type of *Symphyletes*) ; and any reliance on it as a generic character would only tend to separate species which ought to be kept together.

Symphyletes variolosus.

S. subangustatus, fusco-olivaceus, leviter pubescens ; elytris apice sinuatis, bidentatis, fulvo maculatis.

Hab. Australia (Melbourne, Moreton Bay, &c.).

Rather narrow, dark olive, shining, with a very thin, scarcely noticeable pubescence, irregularly and coarsely punctured ; head rather small, a deeply impressed line between the eyes ; antennæ longer than the body, a little shorter in the female, brown, ciliated beneath ; prothorax nearly as broad as long, the anterior margin narrower than the posterior, the sides scarcely rounded, the disk slightly sulcated with three indistinct, interrupted, yellowish bands ; scutellum transverse, rounded posteriorly ; elytra broadest at the shoulder, gradually tapering to the apex, which is sinuated with a short process on each side, almost free from pubescence, except the small yellowish tufts which dot their surface ; legs dark olive ; body beneath with the pile pale greyish, slightly clouded with buff. Length 6 lines.

A rather common species in collections, and having apparently a wide geographic range. Its nearest affinity is with *S. albo-cinctus*, Don. ; but, in addition to other characters, it wants the white band at the sides of the elytra. The females of *Symphyletes* appear to have a longitudinal impressed line in the middle of the last abdominal segment.

Abryna [Lamiidæ].

Newman, Entomologist, p. 289.

Abryna pardalis.

A. robusta, grisescente pilosa, maculis plagisque nigris ornata ; scutello tarsisque nigris.

Hab. Ceram.

Pitchy-black, with a short, close, pale-greyish pile, and spots and patches of black ; head mostly black, the cheeks and vertex spotted with greyish, the epistome clothed with rusty hairs ; prothorax subtransverse, with four obtuse tubercles on the disk (1.2.1), the two lateral teeth distinct ; scutellum black ; elytra rather short, broadest at the shoulders, slightly depressed behind the scutellum, a large black patch externally, a little distance from the shoulder, and rather behind the middle another ; antennæ scarcely longer than the body, all the joints from the third to the seventh inclusive ashy-white at the base, the basal joint nearly black ; legs with a greyish pile tinged with black, the tarsi entirely black ; eyes and mandibles dark brown ; body beneath with a sparse dull-ashy pile mottled with black. Length 9 lines.

Some individuals of this species are much darker than others, and the spots more confluent.

Abryna vomicosa.

A. robusta, grisescente pilosa, maculis nigris irrorata ; scutello grisescente ; tarsorum articulis duobus basalibus albis.
Hab. Cambodia.

Pitchy-black, with a short pale-greyish (or inclining to yellow) pile sprinkled with numerous small black spots ; head rather broad in front, the spots irregular and confused ; prothorax subtransverse, with three obtuse tubercles on the disk (2.1), the posterior divided by a deeply impressed longitudinal line, **the two** lateral teeth very distinct ; scutellum greyish ; elytra rather short, **broadest at** the shoulders, slightly depressed behind the scutellum, clothed **with a** pale-greyish pile, slightly mottled with a darker grey, **and thickly** sprinkled with small black spots, which are formed almost **entirely by** the punctures ; antennæ scarcely longer than the body, **the basal joint** greyish, spotted **with black,** the rest black, except the second and bases of the succeeding **ones to the** ninth inclusive which are ashy-white ; eyes and mandibles **dark** brown ; legs greyish, spotted with black, **the tarsi** black, the **two** basal joints white ; body beneath covered with a coarse greyish **pile, the** sides of the **abdomen** spotted with black. Length 10 lines.

The difference **between** this species and the last is greater than might be imagined **from a comparison** of the two descriptions, but it may be rendered more obvious by remarking that, while the spots are **larger in** *A. pardalis*, they have invariably around the puncture, which forms the centre of each, a circle of black pile, and that these spots often become confluent, having **a more or** less patchy appearance ; **but in** *A. vomicosa* the spots are confined chiefly to the punctures, which then almost **entirely constitute the spots ; the two** basal joints of the tarsi, nearly of **a pure white, offer a remarkable contrast** to the deep black of the remainder.

From *Abryna*, as originally proposed by Mr. Newman, I think it will be necessary to separate those species which approach *Dorcadion* in form and, except very partially in **one or two of** them, in the total absence **of** pubescence. For these I propose the term "*Aprophata*," with the following characters :—

APROPHATA.

Head rounded, **not** dilated below the eyes **in** the male, the vertex and front very convex. Eyes deeply emarginate. Antennæ scarcely longer than the body, not arising from tubercles, the basal joint short, slightly incrassated upwards, the third joint longest, the fourth nearly as long, the remainder shorter and subequal. Prothorax more or less quadrate.

Elytra short, ovate. Legs short, robust. Prosternum slightly produced
posteriorly; mesosternum with a corresponding process anteriorly. External angle of the anterior cotyloid cavities very large.

The principal points which distinguish *Abryna* from *Aprophata*
are the rounded head, especially convex in front and on the vertex,
the ovate elytra, and the large angulation of the anterior cotyloid
cavities; but the habit is so very distinct that it would be doing
violence to all our ordinary notions of generic identity to keep them
together, although, it must be confessed, the technical characters are
not very important, and perhaps rather questions of degree. I have,
however, repeatedly compared all the species one with another, and
I find every character, so far as they can be ascertained without
dissection, usually considered of generic importance, and not mentioned above, more or less variable. The species of *Aprophata* are
excessively rare in collections, very little known, are natives of the
Philippine Islands, and have all been described by Mr. Newman in
a work which is now very scarce (the ' Entomologist '). The following dignoses of the three species may therefore be useful :—

Aprophata eximia. *A.* viridi-metallica, nitidissima; prothorace elytrisque
    maculis piligeris griseis ornatis.
Aprophata fausta. *A.* nigro-chalybeata, nitidissima; elytris cyaneo-me-
    tallicis, immaculatis.
Aprophata notha. *A.* nigra, subnitida; sternorum lateribus abdominisque
    segmento basali margine hirsutis, ferrugineo-fulvis.

The last species has sometimes a slightly purplish tint, and has
been recently received from Manilla (*viâ* Germany), ticketed "*Doliops*,
n. s." In this species, too, the two prothoracic tubercles are wanting.

METON [Lamiidæ].

Head subquadrate in front. Antennæ setaceous, longer than the body,
    arising from two diverging tubercles, the basal joint rather short, gra-
    dually thicker towards the apex, the third and fourth equal and longest,
    the rest more or less equal. Eyes small, deeply emarginate. Lip nar-
    rower than the epistome. Palpi small, slender, the terminal joint
    elongate-ovate. Prothorax nearly equal in length and breadth; a
    short, strong tooth at the side, with small tubercles above. Elytra
    wider than the prothorax, the sides subparallel, the base more or less
    crested. Legs robust; tibiæ clavate; tarsi straight, the distal end
    thickened and covered with short hairs; tarsi narrow, the joints trans-
    verse, except the basal of the intermediate and posterior, which are
    triangular; claw-joint moderate; pro- and mesosterna simple.

I described two species of this genus (but without characterizing
the genus itself) in the ' Trans. Ent. Soc.,' 2nd ser. v. p. 42 (July

1859). It seems to be most allied to *Monohammus* and *Dysthæta*; from the former it is distinguished by the terminal antennary joint not being longer than the one preceding (in ♂), as well as by difference of habit, while *Dysthæta*, Pasc., differs from both in the form of the basal joint of the antennæ.

MONOHAMMUS.

Serville, Ann. de Soc. Ent. de Fr. iv. p. 91.

Sect. 1. Pedes anteriores maris elongatæ.

Monohammus Hector.

M. fuscus, griseo-pubescens, fulvo varius; **prothorace** *lateribus tumido, tuberculo minuto instructo; elytris fulvo irroratis, singulis macula nigra pone medio.*
Hab. Ceram.

Dark **brown**, covered with a fine greyish pile, varied with fulvous; head narrow, elongate, with a deeply impressed longitudinal line extending from the epistome to the prothorax; eyes large; **antennæ more** than three times as **long as the** body, arising from two approximate nearly erect tubercles; lip and epistome short; prothorax about equal in length and breadth, narrowed anteriorly, swelling out considerably **at the** side, and armed with **a small but very** distinct tubercle, the disk with a slightly impressed longitudinal **line**; scutellum rounded posteriorly, hairy, the **centre glabrous; elytra rather** elongate, subtrigonate, **rounded at the apex, granulated at the** base, indistinctly punctured, **-prinkled with fulvous, behind the middle a small** black spot on each; body beneath **dull brown; legs elongate, especially the anterior pair,** which have also **their tibiæ serrated internally, and armed near the** extremity with **a short spine, the two basal joints of the tarsi of the** same pair dilated at the sides. **Length 17 lines.**

The above description is drawn up from a remarkably fine male, with the antennæ alone four and a quarter inches long. The female has a smaller prothorax, nearly parallel elytra, shorter legs, and antennæ not more than half as long again as the body. It is allied to *M. Alcanor*, Newm., *bipunctatus*, Schön., and *fulvo-irroratus*, Blount, all of which are referable to M. J. Thomson's *Rhamses*, a genus which I have not adopted, inasmuch as the single character which separates it from *Monohammus*—the spined protibiæ of the male—is so graduated that in some species, *plorator*, *Antenor*, &c. for example, it is difficult to decide if the little callus, which represents the spine, is sufficient to constitute it a *Rhamses*. The habit, too, is just as variable as in *Monohammus*.

CEREOPSIUS [Lamiidæ].

Head narrow, quadrate in front. Antennæ longer than the body, setaceous,
arising from two approximate tubercles, the basal joint elongate, nearly
cylindrical, the third longest, the remainder gradually decreasing to the
tenth, the eleventh as long, or a little longer. Eyes deeply emarginate.
palpi slender, the last joint elongate, ovate. Prothorax small, transverse, narrow in front, gradually expanding into a strong spine, at the
side, near the base; the spines more or less connected by a transverse
ridge, contracted at the base. Elytra trigonate, convex. Legs short;
tarsi narrow, the basal joint scarcely longer than the second, claw-joint
elongate. Prosternum simple, slightly compressed; mesosternum produced anteriorly.

Cereopsius was a MS. name in use at the British Museum, and
adopted by me a few years ago, but has not been published to the
present time. The genus is allied to *Monohammus*, differing from it,
however, in many characters, as the approximate antennæ, elongate
and nearly cylindrical basal joint, the terminal joint also scarcely
longer than the preceding one, in the form of the prothorax the
whole side swelling out to form the spine, which is placed behind
the middle, the trigonate elytra broadest at the base and rapidly
receding towards the apex, and, lastly, the shorter legs.

To *Cereopsius* must be referred the following *Monohammi* of Newm.:
M. Elpenor, *M. Quæstor*, and *M. Lictor*. The first of these is a nearly
unicolorous form of *M. Prætorius*, Erich. One of the handsomest of
the species has been figured by Mr. White, in the 'Proc. Zool. Soc.' for
1858 (pl. 53. f. 7), under the name of *Cereopsius Helena*. *C. exoletus*,
C. marmoreus, *C. patronus*, and *C. histrio* have been described by me
in the 'Trans. Ent. Soc.' (2nd ser. iv. and v.). It will be necessary,
however, to form a new genus for the latter.

IMANTOCERA [Lamiidæ].

J. Thomson, Arch. Entom. i. p. 188.

Imantocera arenosa.

I. fusca, fulvescente adspersa; prothorace subtransverso; antennarum articulis septem ultimis unicoloribus.
Hab. Cambodia.

Pubescent, dark brown, sprinkled above with pale-fulvous more or
less confluent spots; head with a deeply impressed line between the
eyes; antennæ about one-third longer than the body in ♂, shorter in
♀, the basal joint naked, robust, roughly punctured, the rest with a
pale-fulvous pubescence, the third and fourth joints in both sexes
dilated at the apex, with a thick tuft of hairs, confined to the upper

(*i. e.* when the antennæ are projected forward) and apical half of the latter; prothorax scarcely as long as broad, with three or **four** short irregular transverse grooves, and strongly spined at the side; scutellum triangular, rounded at the sides; elytra not broader than the prothorax (including the spines), slightly round at the side, a large fulvescent patch at the apex, the crest at the base with a row of closely set black granules; legs with **a** pale pubescence, femora dark brown, tibiæ reddish brown; tarsi covered with a short pale-yellowish pile; body beneath dark pitchy-brown, almost naked, with fulvous spots on the metasternum, and a double row on **the** abdominal segments. Length 7 lines.

The three species of *Imantocera* known to me have a strong general resemblance, but, I think, may be easily distinguished by the following characters, which I have tabled together:—

Prothorax short, rather broader than long; basal joint of antennæ naked, or nearly so, rugosely punctate.
Last seven joints of antennæ annulated with black and grey.

I. penicillata, Hope.

Last seven joints of antennæ entirely pale fulvous.

I. arenosa, Pasc.

Prothorax very decidedly longer than broad; basal joint of antennæ pubescent. *I. plumosa*, Ol.

M. J. Thomson's " *Imantocera plumosa*, Hope? (*penicillata*, **White?**)," may be, from the " *elytra paulum abbreviata*," *I. penicillata*, Hope, only that the body beneath is not pilose.

A genus of the Baron Dejean's allied to *Gnoma*, Fab., but I believe not yet described, is *Psectrocera*, the type of which, under the name of *Gnoma? plumigera*, has been figured by Professor Westwood in his 'Oriental **Entomology'** (pl. 5. fig. 3). It has the following characters:—

Psectrocera [Lamiidæ].

Head not broader than the prothorax, elongate behind the eyes. Antennæ longer than the body, the basal joint pyriform, the third, fourth, and fifth elongate, each bearing a tuft of hairs at the apex. Eyes widely emarginate. Prothorax narrowly elongate, the sides subparallel, unarmed. Elytra short, depressed, slightly crested at the base, the crest granuliferous, the apex rounded, entire. Anterior legs longer than the others; tibiæ of the intermediate pair toothed externally; pro- and mesosterna simple.

There is a second species in Mr. Bowring's collection.

PALIMNA [Lamiidæ].

Head moderately broad, quadrate in front. Eyes widely emarginate. Antennæ in ♂ twice as long as the body, distant at the base, arising from two short tubercles, eleven-jointed, the basal joint short, sub-conical, the third twice as long, straight, the fourth shorter, and with the remainder, except the eleventh, subequal. Palpi slender. Pro-thorax irregular, subquadrate, not broader than the head. Elytra more or less subtrigonate, convex, irregular, much broader than the prothorax. Legs robust, the anterior pair in ♂ elongate, and protibiæ curved ; tarsi short, the two intermediate joints dilated, the claw-joint large ; pro-and mesosterna simple, the latter dilated posteriorly.

The type of this genus is *Golsinda tessellata*, Pasc. (Trans. Ent. Soc. 1857, p. 49). At that time *Golsinda* was a MS. name of M. Blanchard's, but recently M. J. Thomson (' Essai Ceramb.' p. 341) has published it with *Golsinda corallina* (White) as the type. But the latter is not congeneric with the species described by me, and hence it becomes necessary to give the former a new generic name. The differences between the two genera are, that in *Golsinda*, Thoms., the basal joint of the antennæ is elongate—as long as the third, in fact—and club-shaped, while in this it is short—not more than half the length of the third—and subconical ; the mesosternum in the former is produced anteriorly and bilobed behind, in the latter it is dilated behind, and not produced anteriorly ; there are also secondary characters in connexion with the antennæ, prothorax, habit, and coloration. Olivier has given a figure of a female of a species of this genus (*Cerambyx annulatus*, 67, t. 20. f. 151), and described the male, which M. Chevrolat is disposed to think may be identical with *P. tessellata*. I have, however, long been of opinion that Olivier's insect represented another species more nearly allied to, or perhaps identical with, one in the British Museum labelled " *Golsinda reticu-lata*," White : this agrees in some respects better with the figure ; and both are from India, while the species described by me has only been received from Borneo. Another species is described in the Entomo-logical Society's ' Transactions,' v. p. 41 (*P. infausta*).

CACIA [Lamiidæ].

Newman, The Entomologist, p. 290.

Cacia histrionica.

C. atra, pubescens ; capite prothoraceque lineis tribus, elytris scutellum versus, et fasciis duabus apicalibus albis.

Hab. Ceram.

Black, sparsely pubescent, punctured ; head with an elevated line from the inner angle of the eye to the epistome, cheeks, front, and vertex white; prothorax nearly quadrate, scarcely wider than the head, white, with two black stripes on each side; scutellum transverse, rounded behind; elytra rather short, a large trilobed patch common to both at the base, an irregular band at the middle, and two others at the apex, which are more or less connected, white; legs black, the tibiæ obscurely ringed with white in the middle, tarsi with the two basal joints white; antennæ hairy beneath to the fourth joint, the fifth very slightly so, black, the second and third joints at the base and nearly the whole of the fourth white; sternum white; abdomen black beneath. Length 6 lines.

This, so far as the proportions between the two colours are concerned, is a very variable species; it is allied to *C. anthriboides* (see *ante*, p. 130).

Eris [Lamiidæ].
Pascoe, Trans. Ent. Soc. 2 ser. iv. p. 110.

Eris annulicornis.

E. brunnea, griseo pubescens, nigro variegatus; elytris sublatis, griseis, antice maculatis, postice subreticulatis; antennis, basi excepta, totis annulatis.

Hab. Cambodia.

Light brown, covered with a short, close, pale-greyish pile, varied with black; head nearly quadrate in front, pale grey with three glabrous vertical lines, the central one becoming impressed between the eyes, two black spots above the epistome; eyes black, reniform; antennæ longer than the body, black, all the joints except the two basal pale ashy at the base; mandibles black; palpi reddish at the tips; prothorax scarcely transverse, a little narrowed anteriorly, the sides smoky-black, continuous with a black patch behind the eye; scutellum transversely triangular, black, the centre and apex pale grey; elytra moderately wide, subbicostate, indistinctly punctured, pale greyish, towards the base a few black spots, behind the middle a sub-reticulate black band, and near the apex an irregular transverse line, also black, more or less connected with small spots behind it, faint spots or mark of a pale leaden grey are also more or less mixed with the black; legs rather robust, femora greyish, with a black band near the apex, tibiæ black, the proximal end and middle grey, tarsi black, with the two basal joints more or less white; body beneath grey, the sides with a few black spots, middle of the abdominal segments glabrous, shining black. Length 7 lines.

Broader and generally more robust than *E. anthriboides*, the colours clearer and more defined, the elytra varied with black, and all the joints of the antennæ, except the two basal, ringed with ashy at the base.

PRAONETHA [Lamiidæ].

Blanchard, Voy. au Pôle Sud, iv. p. 292 (*Prioneta*).

Praonetha subfasciata.

P. breviter subcylindrica, fusca, sparse fulvo pubescens; prothorace subtransverso; elytris medio fascia lata grisea (fere obsoleta) instructis.
Hab. Cambodia.

Shortly subcylindrical, brown, thinly covered with short fulvous
hairs; head convex in front, scarcely as broad as the prothorax; prothorax subtransverse, the anterior and posterior margins nearly equal,
the sides rounded, sparingly punctured; scutellum rather broad,
rounded behind; elytra short, subparallel, irregularly punctured, with
a few black shining granules, principally at the base and along the
suture, a broad but obscurely defined greyish band occupying the middle
third; legs robust; antennæ longer than the body, pubescent, the
basal joint opaque brown, nearly glabrous; body beneath reddish
brown, slightly pubescent, second abdominal segment densely covered
with short hairs at the sides. Length 5 lines.

In many species of this genus the pile is so thin that the derm is
seen beneath, thus producing an obscureness and intermixture of
colours very difficult to define; the broad although somewhat indistinct band, however, occupying just the middle third of the
elytra, in conjunction with its subtransverse prothorax and more
cylindrical form, will readily distinguish this species. As in
Symphyletes pubiventris (ante, p. 339), one of the sexes—probably
the male—has the second abdominal segment densely covered
with short hairs. *Praonetha*, Blanch., is only distinguished from
Pterolophia, Newm., by the absence of the crest at the base of the
elytra—a very slight character, which, as is expressly stated by
Newman, "is sometimes scarcely apparent." *Pterolophia*, however,
appears to me to embrace two forms:—the typical one, including
bigibbera, *varia*, *dispersa**, &c., which are robust, middle-sized insects (5–9 lines), somewhat cylindrical or even compressed; and less
robust and smaller species (2–2½ lines) and as decidedly depressed.
For the latter I have already proposed the genus *Ropica*. Of course
there is nothing satisfactory in such characters when used for the
purpose of generic distinction; but in this and in many other cases
it is doubtful if any more important ones can be found, capable of
embracing a large, or even moderate, number of species. With
regard to those names I do not propose any change here: *Pterolo-*

---

* The two latter were described by me in the 'Ent. Trans.,' under the generic
name of *Notolophia*. I believe there is no such genus: it seems to have been a
slip of the pen for *Pterolophia*. *Prioneta* is probably a typographical error.

phia, although the oldest (1842), is almost unknown to continental entomologists, while *Praonetha* (1853) seems to be generally adopted, nor am I sure that *Pterolophia* has not been already used. It is always more difficult to suppress an old genus than to establish a new one.

Praonetha undulata.

P. olivaceo-brunnea; capite prothoraceque griseo pubescentibus; elytris subunicostatis, apice truncatis, plaga magna ante medium fasciaque dentata apicem versus albescentibus.

Hab. Moluccas (Batchian).

Pale olive-brown; the head and prothorax sparingly punctured, and covered with a short thin greyish pubescence, the latter about equal in length and breadth; scutellum transversely subcordate; elytra slightly compressed posteriorly, a broad but slightly elevated carina near the shoulder, with a very thin greyish pile principally at the base and sides, a large oblique patch before the middle and a very irregular zigzag band behind it white; mandibles glossy black; eyes brown; antennæ scarcely so long as the body. Length 8 lines.

Near *P. albosignata*, Bl., and, after that, the largest of the genus.

Praonetha costalis.

P. rufo-brunnea, pube grisescente varia; elytris tricostatis, costa interiore basi elevata, apice truncatis.

Hab. Batchian.

Pale reddish brown, varied with a greyish pubescence; head and prothorax yellowish grey, with small punctures and patches of brown, the latter nearly quadrate, with the sides slightly rounded; scutellum transverse, rounded behind; elytra subtrigonate, irregularly and rather sparingly punctured, tricostate, the innermost costa elevated or forming a slight crest at the base, the intermediate one less prominent than the inner or outer, the apex truncate, greyish, a broad but indistinct rufous brown band in the middle; antennæ longer than the body, rufous brown; eyes brown; legs obscurely varied with greyish; body beneath dull rufous brown. Length 5 lines.

The abdomen, in my example, is exceedingly small and contracted.

Praonetha penicillata.

P. pallide brunnea, obscure griseo varia; elytris basi subcristatis, postice fasciculatis, apice rotundatis.

Hab. Cambodia.

Pale brown, obscurely varied or clouded with grey; head and prothorax of a nearly uniform grey, finely punctured, the latter subquadrate; scutellum slightly transverse, rounded behind; elytra subtrigonate, seriate-punctate, slightly crested at the base, the apex rounded,

obscurely clouded with greyish, the suture pitchy-brown, a short horizontal tuft of palish hairs posteriorly at the point where the declivity
towards the apex commences, and below this tuft a short curved pilose
line; antennæ a little longer than the body, obscurely ringed with
grey; legs and body beneath indefinitely grey.   Length 5 lines.

One of M. Mouhot's discoveries, easily distinguishable by the
little horizontal tuft on each of its elytra; in some specimens the
suture is unicolorous, or there is a dark-coloured patch at the side of
the elytra.

Praonetha ligata.

P. fusca, pubescens; prothorace subelongato, antice angustiore; elytris
confertim punctatis, apice rotundatis, postice obscure griseo sub-
fasciatis.

Hab. Java.

Dark brown, with a scanty greyish pubescence; head rather narrow;
eyes dark brown; antennæ not so long as the body, obscurely ringed
with grey; prothorax longer than broad, the anterior border much
narrower than the posterior, brown, with two greyish stripes on the
disk; scutellum transversely triangular; elytra slightly narrowing
from the base, covered with large, deep, irregular punctures, a triangular greyish patch indistinct anteriorly, its posterior edge marking
the flexure of the declivity towards the apex, and barely meeting at
the suture, the two together forming an imperfect band; legs and body
beneath pale pitchy-brown, covered with a close grey pile.   Length
6 lines.

I received this species from M. Deyrolle under the above MS.
name, by which, I believe, it is known in the Paris collections.

TRACHYSTOLA [Lamiidæ].
(Dejean), Cat. de Coléoptères.

Head moderate, slightly dilated below the eyes.   Antennæ not longer
than the body, arising from two short remote tubercles, the basal joint
massive, gradually thickened upwards, the rest slender terete, the third
longer.   Eyes deeply emarginate, approximating on the vertex.   Labrum and epistome very short, transverse.   Palpi slender.   Prothorax
transverse, irregular, spined at the side.   Elytra rugose, broader than
the prothorax, sloping posteriorly.   Winged.   Legs rather slender, tarsi
slightly dilated, the basal joint short.   Prosternum simple.   Mesosternum with a vertical tooth.

Dejean places this genus between his *Chæromorpha* and *Penthea*:
the former I do not know, nor has it been published, so far as I am
aware: but its affinity to *Penthea* is by no means evident: it seems
to me better placed near *Dorcadida* and *Microtragus*.

Trachystola granulata.

T. nigra, tota pube fusco-ferruginea induta; elytris seriato-granulatis, seriebus duabus regione scutellari abbreviatis.

Hab. Borneo.

Black, opake, everywhere covered with a short, dense, brownish-ferruginous pubescence; head neither punctured nor sulcated in front; prothorax transversely channeled anteriorly, five flattish tubercles on the disc, arranged ∴, a stout spine at the side; scutellum transversely subcordate; elytra a little depressed on the basal two-thirds of their length, rapidly sloping beyond to the apex, on each nine rows of shining black granules, the inner row distant from the suture, and its granules oblong or almost linear, near the scutellum six granules in pairs, the second row of granules from the suture extending to half the length of the elytra, space between the suture and inner row with two irregular lines of impressed punctures, nearly all the granules with a deep puncture behind. Length 11 lines.

This species differs from a Java congener in the British Museum, labelled *Trachystola scabripennis* (Dej.), in the smaller punctures along the sutural margin, in the second row of granules extending to at least half the length of the elytra, instead of only a quarter, and the double row near the scutellum, whilst there are only two or three altogether in *T. scabripennis.* A third species from Borneo, also closely allied, is in the same museum.

BRIMUS [Lamiidæ].

Head nearly as broad as the prothorax, quadrate in front. Antennæ longer than the body, setaceous, arising from short tubercles, distant at the base, the first joint massive, subcylindrical, the third as long as the first, the rest subequal. Eyes lateral, widely emarginate. Palpi slender, the terminal joint ovate. Prothorax subquadrate, strongly spined at the side. Elytra connate, tapering towards the apex in the male, ovate in the female, the base spined. Legs moderately long, anterior and intermediate coxæ remote, femora subclavate, tibiæ spined, tarsi short. Prosternum simple, mesosternum truncate posteriorly.

Proposed for the reception of *Dorcadion ? spinipenne* (Trans. Ent. Soc. 2 ser. iv. p. 252), which I described from a female specimen in the collection of W. W. Saunders, Esq. There are now five examples in the British Museum, and from one of them, a male, I have drawn up the above characters. *Brimus* differs from *Dorcadion* (to which I doubtfully referred it) in the presence of antennary tubercles and the greater length of the mesothorax, so that the anterior and middle coxæ (as well also the posterior) are separated from each other by a considerable interval, not crowded together so as to be almost in con-

tact : the latter character separates it from *Phrissoma*, from which
it is also distinguished by its non-ventricose elytra and the absence of
all irregularities of surface, except at the base. *Aconodes*, Pasc.,
to which it is nearly allied, has the basal joint of its antennæ short
and fusiform, and scarcely more than half as long as the third. Lastly,
Brimus has a habit of its own distinct from all the rest of the *Dor-
cadioninæ*, although the female has a certain resemblance to Mr.
White's genus *Dorcadida*.

Brimus spinipennis.   (Pl. XVII. fig. 5.)

ATHEMISTUS [Lamiidæ].
Pascoe, Trans. Ent. Soc., 2 ser. v. p. 49.

Athemistus pubescens.

A. tuberculatus, pubescens, rufo-fuscus, setosus ; elytris pone humeros
incurvatis.
Hab. Australia (Port Philip).

Rather narrower than *A. rugosula,* covered above with a dense reddish-brown pubescence, and with longer slender erect hairs interspersed ;
head very convex in front ; prothorax nearly round, coarsely punctured,
a small tooth at the side, and a tubercle above it ; scutellum very small,
triangular ; elytra covered with numerous irregular granulations, narrowly ovate, very slightly prominent at the shoulder, and rather concave behind it, the apex entire ; legs moderately robust ; body beneath
reddish brown, slightly pubescent.   Length 5 lines.

Resembles *A. rugosulus,* Guér. (Parmena), but is at once distinguished by its pubescence.   In Major Parry's collection.

ECHTHISTATUS [Lamiidæ].

Head convex in front ; eyes oblong, scarcely emarginate. Antennæ
setaceous, longer than the body, arising from two diverging tubercles,
the basal joint robust and longest, the third with the remainder subequal. Epistome and labrum small, narrow. Palpi slender, the last
joint obliquely truncate. Prothorax transverse, strongly spined at the
side. Elytra short, ovato-conical, each with a nearly central elevated
spine, the humeral angle extending beyond the base of the prothorax.
Legs long, robust, femora not clavate. Tarsi with the basal joint
nearly as long as the two next together. Prosternum toothed.

The characters which distinguish this genus from *Cerægidion* con-
sist principally in the diverging antenniferous tubercles contrasted
with the remarkably erect and nearly contiguous ones of the latter,
in the toothed prosternum, and the long antennæ, all the joints of
which, except the second, are nearly of equal length ; while in *Ceræ-*

gidion they are scarcely longer than the body, the basal joint being shorter than the third and fourth, which are nearly equal, and the remainder rapidly diminishing. The habit, however, is so similar to *Ceraegidion*, and is in itself so remarkable, that it would be naturally inferred that they were not only nearly allied—as in truth they are—but that they were also natives of the same regions. This supposition is, however, doubtful,—Major Parry, to whom this, I believe, unique Longicorn belongs, having a note to the effect that it was taken from a box of Mexican insects. Notwithstanding, I cannot help thinking that, like *Ceraegidion*, it is a native of Australia.

Echthistatus spinosus. (Pl. XVII. fig. 8.)

E. fusco-piceus, sparse pilosus; prothorace disco subquinquespinoso.
Hab. Australia?

Dark pitchy brown, roughly tuberculate above the interstices, with small patches of short fulvous hairs; head with a V-shaped impression above the epistome; prothorax wider than long, the posterior margin narrowest, the side with a strong median spine, surrounded with tubercles at its base, the disk with short spines, three of which only are at all prominent, two anterior and one posterior, and behind each of the anterior ones two smaller tubercles; scutellum quadrate-cordate, convex, hairy; elytra short, broader than the prothorax at its base, prominent at the shoulder, thence slightly dilating to one-third its length, and gradually rising above into a large somewhat curved spine, then narrowing rapidly to the apex, which is truncate, with the external angle pointed; legs slightly pubescent, the thighs pitchy, tibiæ obscurely ringed with white, tarsi brown; antennæ twice as long as the body; beneath pitchy, with a few dull fulvous hairs. Length 6 lines.

Serixia [Lamiidæ].
Pascoe, Trans. Ent. Soc. 2 ser. iv. p. 45.

Serixia ornata. (Pl. XVII. fig. 9.)

S. rufo-testacea, sat lata; elytris griseo-cervinis, macula communi basali alteraque pone medium albis.
Hab. Moluccas (Batchian).

Rather broad; head reddish testaceous, sparingly pubescent; eyes and mandibles black; prothorax greyish brown, the sides varied with rufous; scutellum transverse; elytra remotely seriate-punctate, pale greyish brown, with a fine silky pubescence, a large and very distinct spot at the base, common to both, and another, on each, behind the middle and towards the side, pure white; antennæ brownish, the third and fourth joints pale at the base; body beneath and legs pale rufous testaceous. Length 4 lines.

This pretty Longicorn connects my *Iolea histrio* with the more

uniformly coloured species represented by *Iolea prolata, longicornis,*
and others ; and at the same time it is so evidently allied to *Serixia,*
that I do not see any characters by which they can be kept apart.
Serixia, as the oldest name, must therefore be adopted.   In addition
to the characters previously given (Trans. Ent. Soc. 2 ser. iv. p. 45),
the genus may also be recognized by the little narrow lobe on the
disk of the prothorax posteriorly, but which never attains to its
margin.

Serixia cephalotes.

S. rufo-testacea; elytris, basi excepta, infuscatis, griseo pubescentibus.
Hab. Batchian.

Moderately narrow, pale reddish testaceous; head and prothorax obsoletely punctured, finely pubescent: scutellum small, triangular ; elytra
remotely seriate-punctate, very dark ashy, and, from the varying light
of the somewhat silky pubescence, much paler in certain positions,
especially towards the apex ; antennæ two or three times as long as the
body, brownish, base of the first and fourth joints testaceous ; legs and
body beneath pale testaceous ; eyes and mandibles black.   Length 3½–4
lines.

In one of my specimens the breadth of the head is nearly twice
that of the prothorax ; in two others it is considerably less, although
still exceeding the ordinary size ; the antennæ, also, are of variable
length.

Serixia sedata.

S. rufo-testacea, sat lata ; elytris grisescente pubescentibus, apice aliquando
infuscatis ; oculis, antennis, mandibulisque nigris.
Hab. Siam.

Reddish testaceous, inclining to ferruginous, with a thin greyish
pubescence ; head and prothorax with shallow scattered punctures, the
latter transverse and narrower than the former; scutellum broadly
triangular ; elytra seriate-punctate, the apex in some individuals black,
more or less brown or entirely concolorous in others ; eyes and mandibles black; antennæ greyish brown, pubescent, half as long again as
the body, rather stout, the fourth joint with the basal half, and occasionally the bases of the sixth and eighth also, reddish ferruginous.
Length 3½ lines.

The rings on the antennæ are in some examples scarcely apparent.

EUMATHES [Lamiidæ].
(Dejean, Cat. de Coléop.)

Head short, narrower below the eyes.   Antennæ setaceous, longer than
the body, distant at the base, the first joint of moderate length and
thickness, the third longest of all, the remainder gradually decreasing

in size. Eyes large, deeply emarginate. Epistome and lip short, the former scarcely broader than the latter; terminal joint of the palpi ovate, pointed. Mandibles entire at the apex. Prothorax subquadrate, slightly toothed at the side. Elytra subdepressed, wider than the prothorax, the sides gradually rounded to the apex. Legs rather short, anterior cotyloid cavity slightly angulated externally, femora subclavate; tibiæ straight; tarsi narrow, the basal joint of the posterior as long as the rest together, the claw-joint short, claws strongly toothed. Prosternum simple, mesosternum keeled.

The toothed claws, combined with the keeled mesosternum, will distinguish this genus from *Hebestola*, which appears to me to be its nearest ally. *Eumathes undatus*, published by me in the Entomological Society's 'Transactions,' 2 ser. iv. p. 251, I believe to be congeneric with *E. jaspidea* (Dej.).

STERNACANTHUS [Cerambycidæ].
Serville, Ann. de Soc. Ent. de Fr. i. p. 172.

Sternacanthus Batesii.

S. ater, nitidus; elytris fasciis subintegris tribus rubris.
Hab. Para.

This insect has long stood in my cabinet as *S. undatus*, Ol. Mr. Bates, however, has recently called my attention to the differences between the two; and since that I have seen two specimens of the true *undatus* in the extensive collection of Wm. Jeakes, Esq., and which were formerly in the possession of the Marquis de la Ferté. In the true *undatus* the bands have precisely the undulating character represented in Olivier's figure, and are very different from the nearly straight, although slightly toothed bands of the *Batesii*; the habit is also different; and were the two insects compared, other characters would doubtless be found to distinguish them.

STENYGRA [Cerambycidæ].
Serville, Ann. de Soc. Ent. de Fr. iii. p. 95.

Stenygra contracta.

S. fusca, nitida; prothorace ampliato, longitudinaliter plicato; elytris medio coarctatis.
Hab. Amazons (Napo).

Dark glossy brown, very sparingly furnished with long pale-yellowish hairs; head moderately elongate, roughly punctured; prothorax subglobose, broader than the elytra, marked longitudinally with numerous fine lines or plaits; scutellum triangular, with a greyish-white pile; elytra elongate, narrowed in the middle, prominent at the shoulders, and raised at the base, the apex rounded and swollen, an oblique narrow yellow basal line, and at about the middle another, but dilated outwardly, the two forming together an interrupted ✕ mark; femora moderately clavate, the posterior with a spine at its extremity, tarsi slightly curved, densely clothed with golden-yellow hairs internally on its lower half, tarsi rather short; antennæ with a silvery pubescence, the terminal joints very strongly dilated. Length 11 lines.

Near *S. coarctata*, Fab., but with the prothorax shorter and more rounded at the sides, its surface longitudinally marked with fine, wavy, more or less connected lines; the elytra much longer and narrower, and more decidedly contracted in the middle, &c. &c.

Sthelenus [Cerambycidæ].
Buquet, Ann. Soc. Ent. de France, 1859, p. 621.

Sthelenus morosus.

S. fuscus, opacus; elytris abbreviatis, singulis maculis elongatis tribus flavis; antennis articulo secundo longiore, incrassato, piloso.
Hab. Caraccas.

Dark brown, opake, with a few stiff black hairs; head wider than the prothorax, and about one-half its length, covered with large, coarse, often confluent punctures, somewhat transversely arranged, especially on the vertex, front slightly concave; lip small, ferruginous; palpi of nearly equal length, pale ferruginous; prothorax nearly cylindrical, a little constricted towards the base, the disk with numerous fine transverse irregular plaits; scutellum rather elongate, rounded behind, somewhat concave; elytra coarsely punctured, much wider than the prothorax, nearly flat above, curved slightly inwards at the side, not extending beyond the base of the fourth abdominal segment, each having three oblong longitudinal patches (the last two nearly continuous) of bright-yellow, curved, appressed hairs; legs rather short, tibiæ and tarsi slender; body beneath pitchy brown; antennæ scarcely longer than the body, the third joint thicker than the basal, and largest of all, hairy, the seventh to the eleventh inclusive short and a little dilated. Length 8 lines.

The above applies exclusively to the male; the female is smaller, more ferruginous, with *longer* antennæ, the terminal joints not dilated, but the third as thick in proportion as in the male. Instead of referring this species to the genus *Sthelenus* of M. Buquet, it will perhaps be thought that it would have been more advisable to have considered it as the type of a new one. I regard *Sthelenus*, however, as very closely connected with *Ozodes*, Lew.; and as in that genus we find the prothorax more or less nodose, and the third (and sometimes the fourth and fifth) joints of the antennæ considerably incrassated, so we may expect to find the same variations in the present. Beyond this, I see nothing to justify its separation from that genus. My example of *Sthelenus ichneumoneus*, Buq., is from the Amazon Valley, and differs in this respect, that the legs are concolorous, except the posterior, which are somewhat darker.

PHORACANTHA [Cerambycidæ].

Newman, Ann. and Mag. Nat. Hist. v. p. 19 (1840).

Phoracantha superans.

P. fuscus: prothorace parvo, subæquali, leviter rugoso, spina laterali elongata recta; elytris elongatis parallelis, pallide fulvis, basi marginibusque castaneis, apice bispinosis.
Hab. Tasmania.

Dark brown; head small, with shallow confluent punctures, an impressed line between the eyes; prothorax small, subequal, covered with coarse confluent punctures, an elliptical space, on the median line near the base, smooth and shining, the side with a slender, elongate, straight spine; scutellum small, triangular; elytra broad, a little depressed, five times as long as the prothorax, pale fulvous yellow, the base and margins dark chestnut-brown, thickly and deeply punctured, gradually decreasing in size and proximity as they approach the apex, each elytron with two smooth elevated lines, not extending to the apex, and terminating in two long acute spines; legs ferruginous, with yellow silky hairs on the tibiæ and tarsi; antennæ ferruginous, covered with a close greyish pubescence, except the basal and second joints, the third to the seventh inclusive armed with a spine at the apex; body beneath pitchy, pubescent. Length 10 lines.

Very distinct, and not to be compared with any other *Phoracantha* that I am acquainted with. The amount of chestnut-brown on the elytra varies.

CERESIUM [Cerambycidæ].

Newman, Entom. p. 322.

Ceresium apiculatum.

C. luteum, subnitidum, punctulatum; elytris lateribus piceis, apice singulatim acuminatis.
Hab. Moluccas (Batchian).

Reddish yellow, shining, nearly free from pubescence, covered with numerous small punctures; head not prolonged in front, a short impressed line between the antennæ; prothorax subparallel, darker at the sides, rather longer than broad; scutellum subcordate; elytra slightly lobed at the shoulder, rather depressed, parallel, the apex of each terminating in a sharp submedian point, irregularly punctured, the sides

darker, inclining to pitchy; antennæ, except the basal joint, lower part
of the tibiæ and tarsi pubescent; mandibles black at the apex; terminal
joint of the maxillary palpi elongate, scarcely triangular, of the labial
narrowly triangular; abdomen impunctate, pale luteous.   Length 4
lines.

CLYTUS [Cerambycidæ].

Fabricius, Syst. Eleuth. tom. ii. p. 345 (1801).

Sect. Antennæ setaceæ, corpore vix longiores.   Prothorax ovatus.
Femora vix clavata.

Clytus patronus.

C. elongatus, subcylindricus, flavo-aurantiacus; prothorace elliptico-ovato;
elytris apice truncatis, externe spinosis, maculis duabus marginalibus
obliquis fasciaque postica atris.

Hab. Batchian.

Elongate, subcylindrical, pubescent, yellowish-orange, light on the
elytra; head nearly vertical, quadrate in front, with a semilunar impression on each side between the eye and epistome; eyes dark brown,
tips of the mandibles black; antennæ setaceous, rather longer than the
body, a little hairy beneath, the fourth joint shorter than either the
third or fifth; prothorax elliptic-ovate, with narrow anterior and
posterior margins; scutellum subtriangular, rounded below; elytra
rather wider than the prothorax, gradually tapering to the apex, which
is truncate, with the outer angle terminating in a spine; each elytron
with three equidistant, black spots, the two first oblique (downwards
and outwards), not connected at the suture, the third forming a continuous band near the apex; legs rather long, slender; body beneath
rather glossy, slightly pubescent.   Length 7 lines.

C. Balyi, Pasc., is the nearest ally of this handsome species.

Sect. Antennæ sublineares, compressæ, corpore vix longiores.   Prothorax globosus, postice constrictus.   Femora haud clavata.

Clytus diophthalmus.

C. rufo-castaneus, sericeus; prothorace maculis duabus nigris; elytris
brevibus, parallelis, integris, dimidio apicali nigris, valde sericeis.

Hab. Queensland (Moreton Bay).

Reddish-chestnut, silky, covered with numerous very fine, erect
hairs; head subtriangular in front; eyes reddish brown, tips of the
mandibles black; antennæ rather long, sublinear, compressed, especially
the terminal joints, the third and fifth of equal length, the fourth much
shorter; prothorax nearly globose, except at the base, where it is
strongly constricted, the disk with a large black spot on each side;
scutellum nearly triangular; elytra rather short, somewhat depressed,
the sides parallel, the humeral angle produced, the apex entire, rounded,
and very convex, posteriorly (but rather less than the half) black, very

silky, the colour varying according to the light, but bordered obliquely in front by a narrow line of straw-yellow, which ascends parallel to the suture for a short distance towards the scutellum; legs long, slender, compressed; body beneath reddish brown, the abdomen black. Length 7 lines.

A handsome and remarkable species, and not to be assimilated to any other known to me. In outline only it may be compared to *C. thoracicus* ; but there the femora are clavate.

Sect. Antennæ subclavæformes, breves. Prothorax globosus.
Femora clavata.

Clytus stenothyreus.

C. niger; prothorace albo maculato ; scutello angustato, albo piloso ; elytris planatis, pubescentibus, marginibus apiceque infuscatis; femoribus rufis.

Hab. Batchian.

Head short, subtriangular in front, roughly punctured, black, with two white hairy stripes between the eyes and antennæ ; eyes rather large, pale fulvous ; antennæ black, subclaviform, half the length of the body; prothorax nearly globose, wider than the head, roughly punctured, black, a line at the side and eleven spots on the disk composed of white hairs; scutellum elongate, narrowly triangular, densely covered with white hairs ; elytra scarcely as broad at the base as the prothorax, then gradually narrowing to the apex, which is truncate, with the outer angle acute, nearly flat above, and, but slightly, bent in at the sides, pale fulvous, darker towards the apex and along the exterior margin, and very sparsely pubescent; femora short, moderately clavate, yellowish red, tibiæ and tarsi dark brown ; body beneath black, the abdomen glossy, with the two basal segments and sides of the metasternum bordered with white. Length 4 lines.

I am unable to compare this well-marked species with any other known to me. The flat elytra very imperfectly covering the abdomen, and in some degree the habit, suggest an affinity, or rather an analogy, with *Stenopterus, Thranius*, &c.

Sect. Antennæ lineares, breves. Prothorax globosus vel subglobosus.

Clytus detcrrens.

C. niger; prothorace brunneo-rufo; elytris macula subbasali fasciisque duabus, una pone medium, altera apicali, albis.

Hab. South Africa (N'Gami).

Head short, transverse in front, black, rather sparsely covered with short white hairs; prothorax globoso-ovate, brownish red, with scattered white hairs ; scutellum small, transverse, rounded behind ; elytra subtruncate at the apex, black, closely covered with short hairs, a round

spot at some distance from the shoulders and towards the side, a band
behind the middle, and another at the apex white; antennæ short,
linear, unarmed; legs of moderate length, femora not clavate; body
beneath black, nearly glabrous, the two basal segments of the abdomen
with a white silky fringe.   Length 4 lines.

This species will rank with the common European forms, particularly such as *C. trifasciatus, ruficornis,* &c.   It is one of Mr.
Anderson the African traveller's captures.

Sect. Antennæ breves, setaceæ.   Prothorax ovatus vel globoso-ovatus.
Femora haud clavata.

Clytus notabilis.

C. elongatus, viridi-flavus; prothorace nigro bimaculato; elytris apice
truncatis, fascia basali literam W simulante, altera media angulata
maculisque posticis duabus ornatis.
Hab. Japan.

Elongate, densely covered with pale-greenish-yellow hairs, and
spotted or marked with black; head small, quadrate in front: eyes,
mandibles, and palpi horn-colour; prothorax ovate, with two black
spots on the disk; scutellum transverse, rounded behind; elytra subparallel, obliquely truncate at the apex, a black V-shaped mark at the
base of each, which, barely meeting below the scutellum, form together
a rude resemblance to the letter W, behind this there is another band
or blotch, zigzag or very strongly toothed, not extending to the side
or meeting at the suture, and midway between the latter and the apex
is a black irregular patch; antennæ setaceous, unarmed, shorter than
the body, black, sparsely clothed with yellowish hairs; legs slender,
elongate, black, with a thin yellowish pubescence, femora not clavate;
body beneath covered with greenish-yellow hairs.   Length 8 lines.

This fine *Clytus* will come into the section that should also contain such species as *annularis, signaticollis,* &c.   I have not
adopted any of the genera of MM. Leconte, Chevrolat, and Thomson,
which they have proposed for comparatively a few of the members
of the old genus *Clytus.*   The species generally comprised under this
name, although remarkably heterogeneous in many respects, are
connected by characters so intermediate, that it appears to me to be
impossible to fix any satisfactory limits to many of these groups.
As an example, the genus *Cyllene,* Newm., confined by M. Thomson,
as I think it should be, to *C. nebulosus,* is by M. Chevrolat (no mean
authority) made to include a number of North American species
also.   Like *Feronia,* which, after having been divided into some
thirty or forty genera by the Baron de Chaudoir, left a large sur-
plusage which could not be placed in any of them, so I believe it

would be with *Clytus*. The genus must be worked out in its entirety, if it is to be divided satisfactorily. Among the *Clyti*, however, there seems to me to be two groups which, by their habit principally, deserve to be distinguished—one *Rhaphuma*, Thoms., including *C. quadricolor*, Lap., *C. leucoscutellatus*, Hope, and *C. placidus*, Pasc., the other, unnamed, comprising *C. lunatus*, Newm., *C. Hardwickii*, White, and *C. cruentatus*, Pasc. Both these groups appear to be well limited and well marked ; but their technical characters, I fear, will not be very valuable.

Zoëdia [Cerambycidæ].

Head subquadrate in front, constricted into a neck behind. Eyes narrow, elongate, deeply emarginated. Antennæ eleven-jointed, filiform, as long as the body ; all the joints, except the second, nearly equal in length, the basal thickened, obconic. Palpi with the last joint narrowly triangular, obliquely truncate. Prothorax nearly as broad as long, narrow anteriorly, a stout tubercle at the side. Elytra broader than the prothorax, subparallel, rounded at the apex. Legs slender; tarsi nearly linear, the basal joint elongate. Pro- and mesosterna simple.

Near *Tillomorpha* and *Euderces*, but differing completely in the form of the head, the prothorax, &c. With *Attodera** it agrees in having the prothorax of a similar character, although more robust, and in its neck ; but the peculiar round, ant-like head of the latter, and its perfectly entire eyes, place it in a different subfamily. Of the two species described below, I have only seen one individual of each.

Zoëdia triangularis. (Pl. XVII. f. 3.)

A. niger, sericeo-pubescens; capite, prothorace elytrisque macula magna triangulari rufis; antennis rufis, articulo basali infuscato.
Hab. Australia (Melbourne).

Head canaliculate in front, finely punctured, and with the prothorax brownish red, inclining to ferruginous; eyes narrow, oblique, slightly emarginate, pale red; antennæ rather shorter than the body, red, the basal joint clouded with brown; palpi and mandibles red; prothorax longer than broad, swelling out considerably at the side behind the middle, where it is as wide as the head, very narrow and produced anteriorly; scutellum triangular, dull brown; elytra wider than the prothorax, especially at the base, the sides parallel, abruptly deflexed and rounded posteriorly, brownish black, with silky and somewhat silvery pubescence, a large triangular reddish patch, the apex com-

* *Pseudocephala*, Newm. This name being preoccupied, as well as a more recent one, *Orthocephalus*, Mr. Thomson informs me (*in litt.*) that he intends to propose *Attodera* in his forthcoming work on the Longicorns.

mencing at the scutellum, and extending downwards and outwards to the side as far as the middle, but drawing up a little as it approaches the suture; legs slender, sparsely clothed with long stiffish hairs, tarsi and lower part of the tibiæ pale ferruginous; sterna and abdomen pitchy black; anterior coxæ very large, contiguous, and greatly exserted. Length 3½ lines.

A single specimen in Mr. Waterhouse's collection.

Zoëdia divisa.   (Pl. XVII. f. 1.)

A. rufo-fulva, sericeo-pubescens; elytris pone medium, femoribusque, basi exceptis, infuscatis, illis medio litura curvata pilosa instructis.
Hab. Australia (Kangaroo Island).

Reddish fulvous, covered with a pale silky or silvery pile; head rather expanded below the eye, canaliculate between the antennæ, which are unicolorous and about as long as the body; eyes slightly emarginate, dark brown; mandibles black at the apex; prothorax scarcely longer than broad, swelling out at the middle into an obtuse knob; scutellum small, triangular, brown; elytra wider than the prothorax, slightly incurved at the side, the apical third brownish grey or pale fuliginous, above which is a darker patch or band, which becomes gradually paler towards the suture, from the side at about the middle a curved line of dark thickly set hairs ascends inwards and upwards, terminating at a distance from the base equalling its own length, and bordered posteriorly by another line of pale yellow; legs dull fulvous, the femora brownish grey, except at the base; body beneath black. Length 4 lines.

A single specimen in Mr. Bakewell's collection.

Mesolita [Cerambycidæ].

Head quadrate. Antennæ setaceous, longer than the body, distant at the base, the first joint short, clavate, the third longest, the rest gradually shorter. Palpi slender, terminal joint of the maxillary ovate. Eyes small, lateral, reniform, widely apart in front. Prothorax ovate, convex. Elytra obovate, swelling out posteriorly, without humeral angles, not larger than the prothorax at the base, the apex divaricate, acute. Pro- and mesosterna simple, continuous (*i. e.* without opposing faces). Legs of moderate length; femora clavate; tarsi gradually dilated to the third joint, the basal as long as the two following, except in the anterior pair.

The absence of humeral angles and the exact apposition of the prothorax to the elytra are the most striking characters of this genus. In these respects it resembles the Dorcadion group among the Lamiidæ; but its affinity appears to be with *Tillomorpha*, *Euderces*, &c., and therefore allied to *Clytus*.

Mesolita transversa. (Pl. XVII. f. 7.)

M. pubescens, fuliginosa, scutello elytrisque basi cinereis, his fascia transversa lunata apicem versus alba.

Hab. Queensland.

Pubescent, smoky brown; head nearly quadrate in front; prothorax broadly ovate, its posterior margin narrowest; scutellum rather large, triangular, cinereous; elytra contracted at the base, gradually swelling out above and at the sides, attaining its greatest size at rather within two-thirds their length, the basal portion dull cinereous, separated from the rest by a sharply angled or zigzag line, a crescent-shaped white band on the posterior third, the apex simply acuminate; legs rather robust; femora slender at the base; the tibiæ slightly ciliated internally, the intermediate and posterior tibiæ also ciliated externally at the apex; body beneath pitchy, with a greyish pubescence; antennæ with the fourth to the seventh joints inclusive cinereous at the base. Length 2½ lines.

Mesolita lineolata.

M. pubescens, fusca, auro lineolata; elytris basi granulatis; antennis ferrugineis.

Hab. Queensland.

Dark brown, pubescent; head in front somewhat transverse, epistome and lip rusty brown, eye bordered with pale-yellowish hairs; prothorax very slightly contracted at the base and anteriorly, its disk with four yellow linear spots (placed ∷), and another on each side; scutellum triangular, dull yellow; elytra contracted at the base, the largest portion at about two-thirds the length, with several fine, golden-yellow, interrupted lines, the apex spined; legs rather robust, femora less clavate, but the tibiæ ciliated as in the last; body beneath pitchy; antennæ and palpi ferruginous. Length 5 lines.

CALLIDIUM [Cerambycidæ].
Fabricius, Syst. Entom. p. 187 (1775).

Callidium inscriptum.

C. testaceum nigro pictum, parce pilosum; prothorace breviter ovato, postice constricto; elytris dense punctatis, nigro fasciatis; femoribus clavatis, ferrugineis, basi pallidis.

Hab. Queensland.

Testaceous, varied with black, with long, scattered, very pale hairs; head short, slightly ferruginous, a darker patch on the vertex; eyes large, black; palpi ferruginous; tips of the mandibles black; antennæ longer than the body, the two basal joints entirely, the remainder at the apex, ferruginous; prothorax shortly ovate, contracted behind, very minutely punctured, a short black dash on each side at the base; scu-

tellum long, subtriangular, black ; elytra wider than the prothorax, the
sides nearly parallel, coarsely and closely punctured, a semicircular
band at the base enclosing the shoulder, a zigzag at the middle, and a
straight narrow band towards the apex black ; legs pale testaceous,
the femora clavate, shining, ferruginous, the base pale ; body beneath
nearly glabrous, brown, darker on the throat and breast. Length 3½
lines.

The nearest ally of this species appears to be —— *signiferum*,
Newm., a much darker and differently marked insect. The latter,
together with *C. scutellare*, Fab. (—— *piceum*, Newm.), is referred to
a genus neither named nor described by that author, but for which Mr.
White has adopted, also without description, the name of *Callidiopis*
(Blanch.). I don't know what the characters may be which are to
distinguish it from the polymorphous *Callidium*. The antennæ and
prothorax are as variable as the coloration.

TMESISTERNUS [Cerambycidæ].
Latreille, Reg. An. v. p. 121 (1829), non Serville (1833).

Tmesisternus exaratus.

T. chalceo-fuscus, griseo maculatus ; prothorace valde transverso ; elytris
    fortiter sulcatis, interstitiis elevatis, fasciis griseis interruptis ornatis,
    apice extus spinosis.

Hab. Aru.

Robust, dark bronze-brown, more or less spotted with patches of
greyish hairs ; head broad in front, narrowed behind the eyes, two
slightly raised lines forming a Λ above the epistome, the vertex canaliculate, four to six spots in a line beneath the eyes, two between and
four behind them ; prothorax very transverse, rounded and narrowed
anteriorly in the male, the border in front nearly straight, dilated anteriorly in the female, and the border broadly emarginate for the reception of the head, dull bronze, coarsely punctured at the side, leaving a
broad, smooth, shining line in the middle ; scutellum transverse,
rounded behind ; elytra strongly sulcated, the interstices forming
broad, raised lines, the central ones more or less united posteriorly and
not reaching the apex, the sulcated lines filled in here and there with
a greyish pile, forming partial spots which assume the appearance of
interrupted bands (two or three—in some individuals scarcely apparent), apex strongly spined externally ; legs and antennæ sparsely
pubescent ; body beneath glossy chestnut-brown, a single white spot
on each side of each abdominal segment. Length 9 lines.

The nearest affinity of this well-marked species is with *S. sulcati-
pennis*, Blanch., from which, amongst other characters, it is distinguished by its metallic colour and apiculate elytra.

Tmesisternus tersus.

T. niger, nitidus, pube subtilissima grisea tectus; elytris disperse punctatis, apice subsinuatis, muticis, fasciis duabus albis ornatis; antennis, tibiis tarsisque rufo-fulvis.

Hab. Goram (Moluccas).

Rather robust, black, shining, the upper surface covered with a uniform, very fine greyish pile, the two bands on the elytra alone have the hairs of a coarser texture; head deeply channeled in front, a nearly straight raised line at the root of the antenna, between them a few punctures only, the vertex scarcely punctured; prothorax transverse, narrowed in front, coarsely punctured on each side, leaving a smooth space in the middle; scutellum nearly round; elytra irregularly punctured, rather convex at the base, the apex subsinuate, unarmed, a pale-greyish or nearly white band at one-third the length of the elytra from the base, another, but curved forwards and narrower, at about the same distance from the apex; antennæ, tibiæ, and tarsi reddish fulvous, sparingly pubescent; body beneath glossy black, the sides covered with a glaucous pubescence. Length 8 lines.

So very closely allied to *T. trivittatus,* Guér., as, except on comparing them side by side, to be readily mistaken for it; besides certain differences of colour, however, *T. tersus* has the punctures on the elytra irregularly dispersed, not forming two or three rows near the suture, and the apex is entirely unarmed; the vertex and front are also very slightly punctured; the clear reddish-yellow colour of the antennæ, tibiæ, and tarsi, and the leaden tint of the rest, contrast strongly with the general olive hue of *T. trivittatus.* I may observe here that Guérin's name is singularly inappropriate; the animal has not *three stripes,* but *two bands.* Boisduval has proposed to remedy this by substituting "*bicinctus;*" but the law of priority, I fear, cannot admit the alteration. Another *Tmesisternus,* from Ceram, although sufficiently distinct at the first glance, appears to me to be only a local subspecies of the present: there is the same general disposition of colours; but the two bands are very indistinct, and the spaces between them and the apex respectively occupied by a series of closely arranged stripes of a pale leaden hue. I have seen a number of both forms, but nothing intermediate. There is still another form, from Makian (a small island near Batchian), so nearly concolorous that, except at the apex, no markings are visible without the aid of a lens.

Tmesisternus herbaceus.

T. fusco-viridis, fusco variegatus; elytris subseriato-punctatis, apice oblique truncatis, singulo macula griseo-alba pone medium externe notato; femoribus tibiisque flavo-viridibus, tarsis rufo-testaceis.

Hab. Mysol.

Rather narrow, shining, dark brownish green varied with brown, with a thin, sparse, greyish pile; head grey, with an impressed line in the middle, and very few punctures on the vertex and front; eyes pale brown; antennæ slender, longer than the body, the basal half pale green, the apices reddish yellow, the remainder darker, brownish, or brownish yellow; prothorax dark green, as long as broad, coarsely punctured, with a smooth median line; scutellum subquadrate; elytra subseriate punctate, one or two faintly raised lines on each, but more strongly marked at the base, the apex obliquely truncate, dark green, behind the middle and close to the external margin a large greyish-white spot, surrounded, but particularly along the side, by dark brown, towards the apex paler, with a brownish indefinite patch; femora and tibiæ pale yellowish green, tarsi reddish testaceous; body beneath glossy chestnut-brown, the sides with a reddish pile. Length 5 lines.

A very distinct species.

Syllitus [Cerambycidæ].
Pascoe, Trans. Ent. Soc. Lond. 2 ser. v. p. 24.

Syllitus Parryi.

S. fusco-niger, obscurus; prothorace antice posticeque rufo; elytris singulis fulvo quadrilineatis, lineis duabus prope suturam conjunctis.
Hab. Australia.

Dull brownish black; head subtriangular, vertex and space between the antennæ black, stripe over the eyes and rest of the head pale reddish, behind the insertion of each antenna a small tubercle; prothorax finely punctured, about half as long again as broad, the anterior half cylindrical, the posterior expanding into a mammiform tuber, and there nearly as wide as the elytra, the base contracted, the disk with four tubercles, the two posterior largest, the anterior and posterior margins pale red; scutellum convex, rounded, brown; elytra narrow, parallel, each with four pale-yellow, raised, smooth, longitudinal lines, the two towards the suture united near the apex, the third about two-thirds the length of the first, the fourth marginal, the spaces between the lines punctured; legs black; pro- and mesosterna and four anterior coxæ red, metasternum and abdomen black, the latter with a silvery pubescence. Length 6 lines.

This species will be at once distinguished from *S. rectus, grammicus,* and *deustus,* not only by its greater size and more robust form, but by its quadrilineated elytra and dark-brown nearly black prothorax. In the fifth volume of the 'Transactions of the Entomological Society,' n. s., I proposed to separate, under the name of *Syllitus,* those species of *Stenoderus* with elevated longitudinal lines on the elytra, from the ordinary red and black ones which constituted the genus originally. The technical characters which distinguish it are perhaps only of secondary importance, as is the case with many others in the Longicorn families, yet taken in connexion with the fact that one has a type of coloration different from the other, will, I think, justify its adoption.

Dæsus [Cerambycidæ].

Head rounded, slightly contracted behind the eyes. Antennæ setaceous, distant at the base, longer than the body, the first joint short tumid, the second very short, the remainder subequal. Eyes very large, oblong, nearly entire. Lip very small, rounded anteriorly. Palpi growing gradually thicker, the last joint subtriangular. Prothorax nearly equal in length and breadth, narrower in front, rounded behind, the sides carinated. Elytra wider than the prothorax, parallel, the humeral angle produced. Legs moderate, tibiæ slightly curved externally, their margins tuberculate and fringed with short hairs, the first tarsal joint shorter than the two next together. Abdomen soft.

The above description is drawn up from what appears to be a male, in the collection of Major Parry. It has a striking resemblance to a *Telephorus*, but is related to *Vesperus*, although the form of the head and prothorax is so far different that we miss the slenderness which gives such a remarkable contour to the species of that genus; the presence also of a well-marked carina along the side of the prothorax, which, however, does not extend its whole length, would alone suffice to distinguish it. It may also be noted that whilst in *Vesperus* the tibiæ are slender and perfectly straight, in *Dæsus* they are tolerably robust and curved externally, and the basal joint of the tarsi is shorter than the two next together, which is not the case in *Vesperus*. The form and position of the coxæ, palpi, and antennæ, except that the latter are more distant at their insertion, are so far identical as to call for no further notice.

Dæsus telephoroides. (Pl. XVII. fig. 4.)

D. testaceo-ferrugineus, subnitidus; elytris breviter pilosis; oculis nigris. *Hab.* India.

Testaceous inclining to ferruginous; head, prothorax, femora, except beneath, and basal joint of the antennæ smooth, somewhat shining; elytra covered with very short greyish hairs, and each with three slightly raised lines; body beneath paler, with a very sparse pubescence; eyes black; tips of the mandibles dark brown; head slightly broader than the prothorax, rather convex in front; eyes prominent; prothorax but slightly convex; elytra considerably wider than the prothorax, rather elongate. Length 7½ lines.

AMIMES [Cerambycidæ].

Head very short and rounded in front, narrowed behind the eyes. Antennæ eleven-jointed, arising between the eyes from short divaricate tubercles, two or three times as long as the body, setaceous, the basal joint short, narrowly subpyriform, the third twice its length, the remainder gradually longer. Eyes large, prominent, reniform. Palpi slender, pointed. Lip and epistome very short and transverse. Mandibles entire at the apex. Prothorax elongate, irregularly subcylindrical, narrower than the head and elytra, unarmed. Elytra subparallel. Legs slender, basal joint of the tarsi elongate. Anterior cotyloid cavity widely angulated externally, open behind; its coxæ conical, approximate. Pro- and mesosterna simple.

The specimen from which the above generic details have been drawn up was originally described by me in the 'Trans. Ent. Society,' 2nd ser. iv. p. 238, as *Psilomerus*? *macilentus*. The generic name was a MS. one used for a congener at the British Museum, but, as I afterwards found from an inspection of the true *Psilomerus* at Paris (Jardin des Plantes), had nothing whatever to do with my species. As I cannot refer it to any published genus, a new name has therefore become necessary. With regard to its affinities, I have with some hesitation placed it near *Methia*, Newm., hitherto forming with *Dysphaga*, Hald., a small group, originally proposed by Leconte, and principally characterized by its anterior cotyloid cavities open behind. M. James Thomson in his 'Essai,' p. 128, combines *Dectes*, Leconte, with them; but this and *Dysphaga* I have not seen.

Amimes macilentus.   (Pl. XVII. fig. 6.)

MACRONES [Cerambycidæ].
Newman, The Entomologist, p. 33.

Macrones acicularis.

M. angustissimus, ferrugineus; elytris rufo-testaceis, unicostatis; tarsis posticis albis.

Hab. Australia (Adelaide).

Very narrow and elongate, ferruginous; head punctured in front, deeply impressed between the antennæ, the vertex dark brown; prothorax punctato-granulate, very irregular, with a protuberance at the side near the base, and another on the disk above it; scutellum small, bluish-black; elytra terminating at the end of the third abdominal segment, reddish testaceous, with a strongly raised longitudinal line on each; abdomen above dark brown; legs slender, posterior tarsi yellowish white; body beneath brown, abdomen at the base ferruginous; antennæ not reaching to the end of the elytra, dull brown, the basal joint ferruginous, the three apical yellowish white. Length 8½ lines.

ACYPHODERES [Cerambycidæ].
Serville, Ann. de la Soc. Ent. de France, ii. p. 549.

Acyphoderes brachialis.

A. fuscus; capite prothoraceque sericeis, fulvo variis; elytris vitta flavescenti; femoribus intermediis posticisque, basin versus, flavo annulatis, tibiis anticis intus dentatis; abdominis segmento penultimo dente bifido instructo.

Hab. Brazil.

Dark brown; head narrow and elongate, a patch of yellow silky hairs between the eyes, dividing into angular branches below them; prothorax ovate, narrower than the elytra at the base, covered with a silky pubescence varied with four rather indistinct yellow stripes; scutellum narrowly triangular, pale yellow; elytra extending to the middle of the third abdominal segment, punctured at the base, a yellow vitreous stripe from near the base to the apex; legs more or less hairy, especially on the inner side of the intermediate and posterior tibiæ; anterior tibiæ with a strong tooth beneath, near the middle; intermediate and posterior femora annulated with yellow towards the base; body beneath dark brown shining, the metasternum varied with indistinct patches of yellow silky hairs, abdomen elongate, very slender, the basal segment narrowest, the penultimate furnished with a broad bifid tooth at its apical margin. Length 9 lines.

The curious bilobed tooth beneath the abdomen is not, I think, a sexual character, as might be supposed, as it also occurs in what appear to be both sexes in one or two other species of this genus. The abdomen is very much attenuated at the base—a character which, in the group to which it belongs, appears to be only of specific importance. I have not seen any other species having the protibiæ toothed.

HESTHESIS [Cerambycidæ].
Newman, Ann. Nat. Hist. v. p. 17 (1840).

Hesthesis plorator.

H. niger; prothorace margine antica, elytrisque macula apicali flavidis; abdomine supra, segmento primo basi tertioque apice, et infra tribus primis flavo marginatis; femoribus rufo-ferrugineis.

Hab. Melbourne.

Black, with patches or lines of pale-yellow hairs; a patch of yellow hairs in the concavity between the eyes: prothorax subtransverse, tumid at the side, closely punctured, the anterior margin bordered with yellow hairs; scutellum black, triangular; elytra greyish brown, lighter at the base, shoulder and an oblique line at the apex covered with yellow hairs; abdomen above with the first segment at the base, margin of the third, and beneath the first three at the apices bordered with yellow hair; legs reddish ferruginous; antennæ black; posterior angle of the metathorax yellow. Length 7 lines.

Differs from *H. mœrens*, Pasc., in the narrower prothorax, longer elytra, the absence of the yellow border at the apex of the first abdominal segment above, and in the first three segments beneath margined with yellow.

DISTICHOCERA [Cerambycidæ ?].
Kirby. Trans. Lin. Soc. xii. p. 471.

Distichocera mutator.

D. ater; prothorace vittis duabus elytrisque rubro-aurantiacis.
Hab. Queensland.

Deep black; two broad lateral stripes on the prothorax, and the elytra, reddish-orange; head produced anteriorly, deeply grooved between the antennæ, a broad longitudinal excavation on each side in front, a silvery pubescence beneath the eyes, which are of a pale horn-colour; prothorax rather broader than long, black, a wide orange stripe on the disk on each side; scutellum triangular, black, bordered with orange; elytra slightly narrowing from the shoulders, the apex sub-truncate, the outer angle toothed, each with five elevated lines, the intervals closely and finely punctured; antennæ about two-thirds the length of the body; legs slender, the tarsi fringed with silvery hairs. Length 9 lines.

This is so exceedingly like the female of *Distichocera maculicollis*, Kirby, that it might be very readily taken to be the male if we had not been already well acquainted with the sex of that species.

EXPLANATION OF THE PLATES.

PLATE XVI.

Fig. 1. *Dæothena platypoda.*
  „ 2. *Ethas carbonarius.*
  „ 3. *Rhypasma pusillum.*
  „ 4. *Aposyla picea.*
  „ 5. *Zygænodes monstrosus.*
  „ 6. *Phenace œdemerina.*
  „ 7. *Ochotyra semiusta.*
  „ 8. *Piœnia saginata.*
  „ 9. *Ino ephippiata.*

PLATE XVII.

Fig. 1. *Zoëdia divisa.*
  „ 2. *Goëphanes luctuosus.*
  „ 3. *Zoëdia triangularis.*
  „ 4. *Dæsus telephoroides.*
  „ 5. *Brimus spinipennis.*
  „ 6. *Amimes macilentus.*
  „ 7. *Mesolita transversa.*
  „ 8. *Echthistatus spinosus.*
  „ 9. *Serixia ornata.*

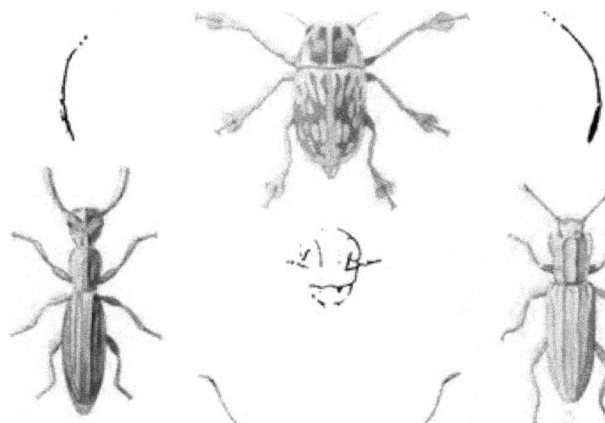

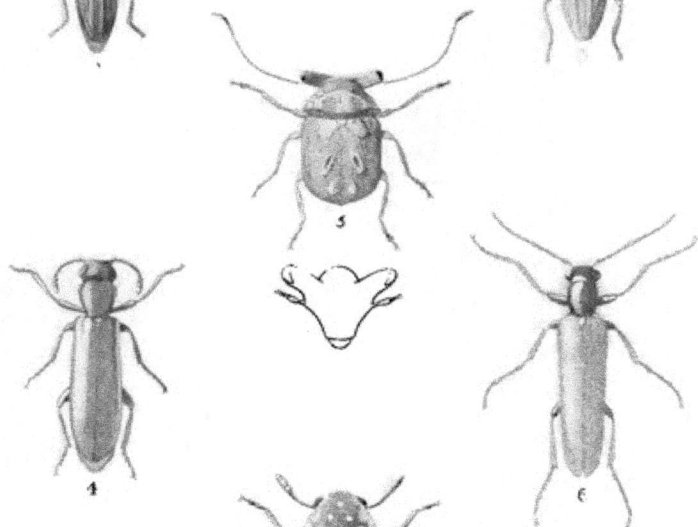

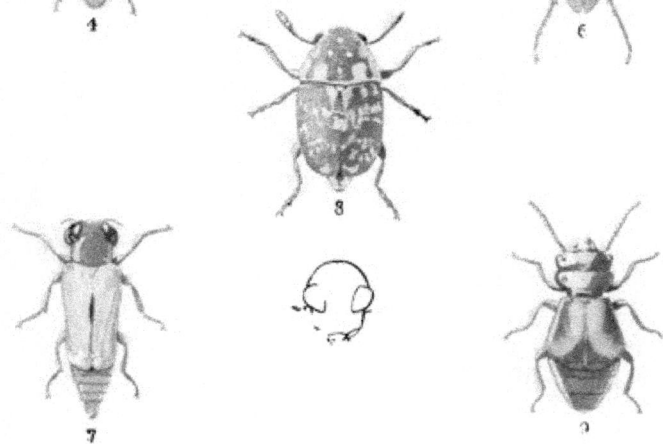

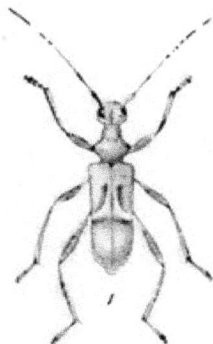

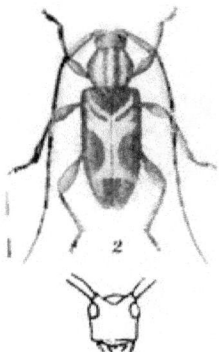

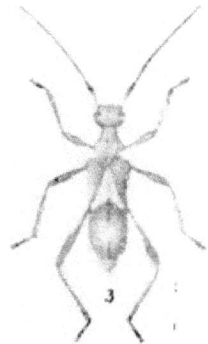

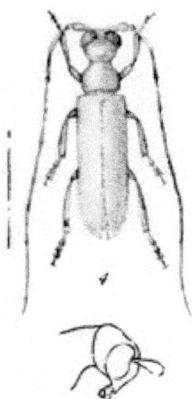

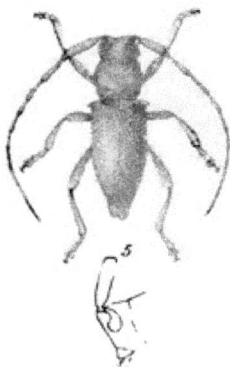

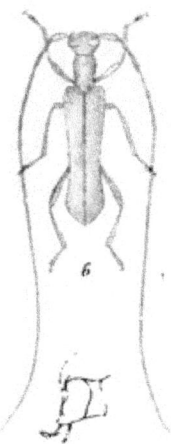

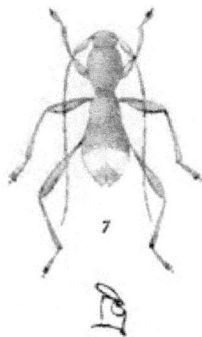

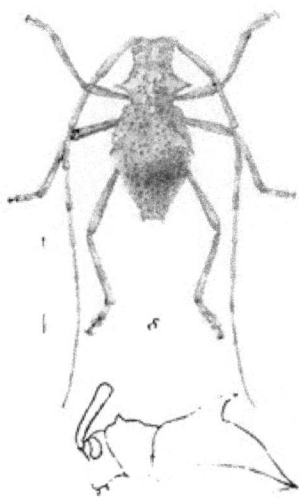

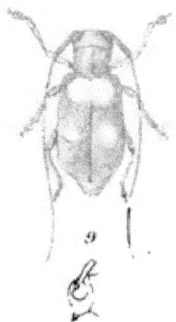

[*From the* JOURNAL OF ENTOMOLOGY, No. VII., 1863.]

ON

CERTAIN ADDITIONS TO THE GENUS DICRANOCEPHALUS,

AND

NOTICES

OF

NEW OR LITTLE-KNOWN GENERA AND SPECIES

OF

COLEOPTERA.

BY

FRANCIS P. PASCOE, F.L.S., &c.

On certain additions to the Genus Dicranocephalus.

The reality of the existence of species has been questioned by many naturalists; not, however, in the Darwinian sense—that is, that as all organic beings have descended from "some one primordial form," they only differ from each other in degree, and, therefore, that classes, orders, families, genera, and species only exist as artificial combinations,—but in the sense of "special creations," and the impossibility of drawing any satisfactory line between species and varieties.

The disbeliever in the material existence of species, however, need not abandon the use of the term: as Agassiz has remarked, "species exist as categories of thought, in the same way as genera, families," &c.; and the only difference between a species and a variety appears to be, that in the first the distinctive characters are more important or more numerous than in the second, and are not bridged over by intermediate gradations, as is frequently observed in the variety. Latterly the word *subspecies* has been adopted to express a grade between species and variety, but at the same time it has been generally connected with, or assumed to be limited to, a certain geographical area. As I take it, the subspecies being dependent for its differential characters on physical, perhaps combined with other causes, and those causes being removed, it would return sooner or later to the normal condition of the species from which it had originally been derived. The species of many genera are, however, so homomorphous, as often to suggest the idea of their having had a common parentage; and no doubt it will be found to be so in many cases where their describers have been but too ready to consider the slightest variation of specific importance*.

These remarks are rendered necessary, because, in the following proposed additions to the genus *Dicranocephalus*†, I do not put forward the three forms described as " undoubted species,"—although it would not be difficult to cite many instances where, in other cases, this has been done on slighter grounds ; nor are they, in the present state of our knowledge, to be considered as merely geographical subspecies, and still less as instances of dimorphism. It is possible, and indeed not unlikely, that intermediate forms may hereafter be received. There is but a moderate gap to be bridged over; but, until that is done, I am sure that it would be contrary to all ordinary notions of specific distinction to unite them under the same name.

The first of these forms, *Dicranocephalus Wallichii*, was brought from Nepaul by General Hardwicke more than thirty years ago, and was described‡ by the late Rev. F. W. Hope in Gray's 'Zoological Miscellany,' afterwards figured§ by Gory and Percheron in their work on the " Cetonides," and later by Prof. Westwood in his 'Arcana.' I believe there were only two representatives of the genus in Europe until Mr. Fortune went to China, when he sent home altogether a large number of specimens, which were, and have continued to be, referred to *D. Wallichii*. Mr. Bowring, however, as I understand, protested from the first at considering it identical with the old species. It is not merely as a compliment, therefore, that I have named it after him.

* A striking instance of this occurs in *Paludomus aculeatus*, a river shell of Ceylon, which, according to Mr. Blanford, in a communication to the Linnean Society, has been split into no less than twenty-four species, all of which he demonstrated, by a large series of specimens exhibited at the meeting, to be reducible to one ! (Trans. vol. xxiii. p. 603.)

† Often erroneously spelt *Dicronocephalus*.

‡ Shortly described, but without a word of a generic kind. *Dicranocephalus* remained a mere catalogue name until the publication of the third volume of the 'Genera des Coléoptères ;' M. Lacordaire must therefore be cited as its authority. MM. Gory and Percheron, in their hybrid jargon, called it " *Goliath Wellech*." *Dicranocephalus* itself is an abominably unwieldy name, and had been previously used by Hahn for a genus of Hemiptera, but it does not appear to have been adopted.

§ The figure is very characteristic, and correctly drawn and coloured. That in the 'Arcana,' from its position, is less satisfactory, and is coloured a pale green. Mr. Hope's phrase is " *pallide flavo-viridis*." I should have thought that the British Museum specimen, from which Professor Westwood took his drawing, might originally have been green, but that the words of the French authors, " *gris-jaunâtre*," are as applicable at this moment as they probably were originally.

Within the last week or two I have been presented with a fine set of Coleoptera collected in Japan, the coast of Mantchouria, south of the Amoor, &c., by Arthur Adams, Esq., late Surgeon of H.M.S. 'Actæon;' and among others there was a specimen of the genus from Chosan, in the Corean Peninsula, which, on comparison, I found to differ from the other two, and this I have dedicated to the generous donor. I will first give the diagnostic characters of the three forms, and then a comparative view of their differences, which will be more intelligible, I think, than a more minute description.

Dicranocephalus Wallichii, Hope.

D. griseo-pubescens; prothorace lato, turgido, carinis duabus, medio, elongatis; elytris breviter subquadratis.
Hab. Nepaul.

Dicranocephalus Bowringii.

D. griseo-pubescens; prothorace modice convexo, carinis duabus, medio, distinctis, brevibus; elytris angulo humerali triangulari-impresso.
Hab. North China.

Dicranocephalus Adamsii.

D. griseo-pubescens; prothorace modice convexo, carinis duabus, medio, parum obsoletis; elytris angulo humerali rotundato, haud impresso.
Hab. Corea.

The first, *D. Wallichii*, is a very much broader insect; the prothorax very convex, and swollen anteriorly, and, if viewed in profile, presenting a very considerable curve,—the two carinæ on its disk distinctly raised, narrowly and strongly defined, and, from the curve, appearing much longer; the elytra are broader and more quadrate, that is, less narrowed behind; and the tarsi are testaceous yellow, ringed with black.

The second, *D. Bowringii*, has also the tarsi coloured in the same way, and the carinæ on the prothorax are equally well defined, although shorter; but the other characters present a marked contrast to the above.

The third, *D. Adamsii*, has the tarsi entirely black; and the carinæ have nearly disappeared, leaving only two broad marks, which gradually shade off on each side, but are tolerably distinct in the middle, owing to the sudden dip of the longitudinal cavity, which is alike common to all the forms; and the humeral angle, which forms a sort of boss, is rounded, and without the triangular cavity of *D. Bowringii*.

These are not all the differences between the two last forms. *D.*

Adamsii has the basal joint of the anterior tarsus not longer than the second, and the intermediate and posterior tibiæ are much shorter than the corresponding parts in *D. Bowringii*; in the former the head is broader within the two lines which extend up the front from the horns, and is without the concavity which distinguishes the other.

Briefly to sum up the most prominent points, *D. Wallichii* is distinguished from the other two by its greater breadth and its turgid prothorax, and *D. Bowringii* from *D. Adamsii* by the triangular impression on the shoulders, always filled in by the pubescence which has escaped the abrasion which is suffered by the more projecting parts.

Notices of new or little-known Genera and Species of Coleoptera.

[Continued from vol. i. p. 371.]

PART IV.

SILPHOMORPHA [Carabidæ].
Westwood, Trans. Linn. Soc. xviii. p. 415.

Silphomorpha speciosa.

S. late ovata, subtilissime punctata, viridi-purpureo-metallica, nitida, subtus nigro-chalybeata; antennis ferrugineis.
Hab. Queensland.

Broadly ovate, very minutely punctured, deep golden green, with brilliant dark purple or violet reflexions; body beneath and legs black, with a chalybeate gloss; femora greenish metallic; antennæ and palpi ferruginous; eyes pale; head finely corrugated and punctured, deep violet, bordered with green in front, the lip black; prothorax very transverse, bisinuate anteriorly, with very minute punctures, and divided by irregular lines into exceedingly fine reticulations; scutellum triangular, black; elytra lightly seriate-punctate, the interspaces also minutely punctured; body beneath finely corrugated, the penultimate abdominal segment deeply emarginate; tarsi dark ferruginous. Length 8 lines.

This magnificent species is very distinct from any other in the remarkable subfamily to which it belongs, but apparently a true *Silphomorpha*. The purple or dark-violet reflexion (it is difficult to fix which colour-name is most appropriate) is more decided at the base and centre of the elytra, and is also very marked at the sides of the prothorax. In my collection, and I believe unique. A coloured figure will be given in a supplemental plate.

CEPHALODESMIUS [Scarabeidæ].
Westwood, Trans. Ent. Soc. 1 ser. iv. p. 117.

Cephalodesmius laticollis.

C. niger, opacus; clypeo antice bidentato, dentibus duobus mediis basi separatis; prothorace elytris latiore.
Hab. Queensland.

Dull black, opake; head very transverse; the clypeus four-toothed, the two central teeth longest, linear, subparallel, and widely separated at the base; prothorax very broad, wider than the elytra, and presenting an almost foliaceous margin at the side anteriorly; elytra obsoletely striated, slightly convex, almost concave towards the shoulder; body beneath and legs dull black; palpi ferruginous. Length 7 lines.

Well distinguished from *Cephalodesmius armiger*, Westw., the only species of this genus hitherto described, by the slight convexity of its upper surface, the breadth of the prothorax, by the direction of the two central teeth of the clypeus, and their separation at the base. The head is also broader and shorter, the legs longer, and the abdomen more contracted.

DIATELIUM [Scaphididæ].

Caput collo elongato; *oculis* magnis, integris, rotundatis. *Antennæ* graciles, clava quinque-articulata. *Palpi* subulati, acuti. *Scutellum* liberum. *Pedes* elongati, *tibiis* bicalcaratis. *Mesosternum* carinatum.

Notwithstanding the extraordinary form of this insect, owing to its exceedingly long neck, it is very closely allied to *Scaphidium*, differing from it principally in that respect and in its entire and prominent eyes. As in the Scaphididæ generally, the abdomen has six segments, and the prothorax and elytra have the same peculiar punctation. Mr. Wallace has taken it both in Sumatra and in Borneo.

Diatelium Wallacei. (Pl. II. fig. 2.)

D. fulva, nitida; capite, prothoracis basi et medio, elytrorum macula discoidea clavaque antennarum nigris.
Hab. Sumatra; Borneo (Sarawak).

Fulvous yellow, very smooth and shining; head and neck nearly as long as the rest of the body together, black, the latter finely corrugated transversely; eyes fulvous; antennæ pale testaceous, short, arising from a round fovea in front of each eye, the last five joints black, forming a loose club; prothorax rounded anteriorly, convex, the middle and base black; scutellum black; elytra rather depressed, a large black discoidal spot on each; body beneath fulvous; coxæ, base and extremity of the femora, the mesothorax, and the episterna of the metathorax, as well as its posterior border, black. Length 6 lines.

CLIDICUS [Scydmænidæ].

Laporte, Ann. de Soc. Ent. de France, i. p. 397.

Clidicus formicarius. (Pl. II. fig. 3.)

C. setulosus, rufo-piceus ; prothorace subcordato ; pedibus piceis.

Hab. Borneo (Sarawak).

Rufous pitchy, covered with short, stiff, erect hairs ; head almost obsoletely punctured, shortly triangular, bilobed behind, an elevated transverse interocular ridge beneath which and at each end arise the antennæ ; eyes very small, round, lateral ; antennæ claviform, the basal joint obconic, as long as the next four together, and more or less triangular, gradually increasing in size to the seventh, the last four shortly transverse ; lip and epistome short, transverse ; mandibles short, curved, glossy black ; maxillary palpi very long, the last joint ovate, pointed, and nearly as long as the preceding, the labial short, the last joint subulate ; prothorax obscurely punctured, subcordate, considerably rounded anteriorly, narrower than the head, to which it is attached by a short neck ; scutellum very small, triangular ; elytra ovate, convex, each with six shallow striæ, which are very coarsely punctured ; anterior coxæ elongate, contiguous, the middle and posterior separated by a slight interval ; femora subclavate ; tibiæ fusiform, unarmed ; tarsi subfiliform, all their joints, except the last, of equal length ; abdominal segments six ; winged (?). Length 3 lines.

To *Clidicus* belongs the genus *Erineus,* Walker. The species described by him (*E. monstrosus*) differs from the above in its subquadrate prothorax and other characters. *Clidicus grandis,* Lap., is a more slender form, with longer legs, antennæ, &c.

NARCISA [Trogositidæ].

Caput insertum, fronte verticali. *Oculi* divisi, superiores remoti, verticales. *Antennæ* breves, articulo primo incrassato, clava subunilaterali triarticulata. *Maxillæ* lobo interiore obsoleto. *Prothorax* transversus, lateribus foliaceis. *Elytra* marginibus subdilatata, serrulata. *Corpus* ovatum, subdepressum.

This genus will be at once distinguished from *Anacypta* by the remoteness of the upper eyes, and the serrulate and partially dilated border of the elytra ; and from *Gymnochila* by the foliaceous sides of the prothorax, and by the less decided unilateral position of the club of the antennæ, as well as by habit.

Narcisa decidua. (Pl. III. fig. 5.)

N. obovata, pallide ferruginea, squamis albidis tecta ; antennis rufescentibus.

Hab. Batchian.

Obovate, pale ferruginous, rather sparsely covered with greyish-white scales; head dark brown, deeply set in the prothorax; eyes black, rather small, vertical, remote; antennæ rufous, the club partially unilateral, with its first two joints very transverse; external maxillary lobe narrow, ciliated, the internal obsolete; maxillary palpi with the terminal joint elongate-ovate, of the labial shortly ovate; prothorax more than twice as broad as long, the sides dilated, their edges with rounded serratures; scutellum transverse, rounded behind; elytra narrower than the prothorax at the base, dilated at the shoulder, then gradually rounded with the margin less and less dilated to the apex, its edges serrated and fringed with setose scales (where the scales have fallen off, the elytra are seen to be crenate-striate, with traces of darker or brownish spots, which form a sort of band, one near the base, the other towards the apex); body beneath dark brown, the legs paler and covered with smaller scales. Length 3½ lines.

LEPERINA [Trogositidæ].

Erichson in Germar, Zeitschr. für die Entom. v. p. 453.

Leperina turbata.

L. late oblonga, aterrima, supra fusco-nigro squamosa, fasciculis nigris plus minus elongatis induta; elytris subparallelis, maculis duabus albis posticis.

Hab. Australia (Sydney?).

Broadly oblong, deep black, rather closely covered with small black and greyish scales, mixed with more or less elongate, erect or semi-erect scaly hairs, generally collected together in fascicles; head and prothorax with coarse scattered punctures, from which the scales arise; these are principally directed forwards, and are mostly greyish, a few only being black; a fascicle of black hair-like scales over each eye, and a large one nearly adjoining on each side the prothorax, on the latter a slightly raised median line; scutellum triangular, fasciculate; elytra nearly parallel, with two strongly marked costæ, three long black fascicles (longitudinally disposed), among many smaller ones, on each, between the middle and apical fascicles a silvery-white patch; legs and body beneath black, coarsely punctured, and sparsely covered with greyish setose scales; the prosternum smooth and polished. Length 6 lines.

In its long scaly fascicles this species resembles *Leperina cirrosa* (*ante*, vol. i. p. 100), but is much larger and proportionally broader, and the white scales are chiefly confined to a single spot on each elytron.

CRINE [Nitidulidæ].

Caput late triangulare, ante antennas sulcatum. *Antennæ* breves, duo-decim-articulatæ; clava ovata, triarticulata. *Palpi* crassi, cylindrici.

Tibiæ trigonatæ. *Tarsi* quatuor- vel quinque-articulati, articulis tribus primis dilatatis, brevissimis. *Corpus* depressum.

The curious little insect constituting this genus belongs to the subfamily *Rhizophaginæ*, hitherto composed of *Rhizophagus* only, but to which I would also refer *Europs*, Woll., and *Nomophlœus** and *Hesperobœnus*, Motsch.† The two latter, however, appear to me to be identical. There are several discrepancies among authors in their descriptions of *Rhizophagus*. In the first place, Erichson denies that there are two lobes to the maxilla, as Curtis had represented; but M. J. du Val says that in this he is most certainly in error. Again, M. Lacordaire allows only ten joints to the antennæ, the ninth and tenth forming the club. M. J. du Val gives eleven; but in the two species which he has figured in his great work ('Coléopt. d'Europe') twelve are represented, as is the case also in Mr. Curtis's plate. As M. J. du Val states, there are unquestionably two lobes to the maxilla; and as unquestionably, I should say, are the antennæ twelve-jointed, as MM. Curtis and Migneaux have represented,—the last forming a little knob on the eleventh; but the two, although minute, are perfectly distinct. Exception may be taken that these are not true articulations, especially the last; but in any case the ninth has nothing to do with the club. They are here described as 12-jointed, as I cannot understand on what principle the last is to be ignored any more than the one preceding it. The line of punctures, which form a sort of oval on the prothorax, recalls the impres-

---

* Whilst these sheets were passing through the press, I have had the opportunity of examining for the first time Dr. Leconte's 'Classification of the Coleoptera of North America.' In this work *Hesperobœnus* and *Nomophlœus* are placed in the new family " Monotomidæ," which is " at once" separated from all Nitidulidæ by the " form of the anterior coxæ " (rounded in the former, transverse in the latter). Under the microscope it *appeared* to me that in some a transverse form was more or less assumed when the leg was thrown backwards; this was the case with the large, apparently rounded coxæ of *Crine*; but in *Europs* they are decidedly transverse. It is only necessary to examine the more recent entomological works (particularly the 'Genera des Coléoptères d'Europe,' *passim*) to see the wide divergence of statements in reference to mere matters of fact, where they concern the minute structures. On this account I hesitate trusting implicitly to these delicate characters, so difficult in most cases to realize.

† I have been unable to procure Colonel Motschulsky's 'Études Entomologiques,' in which, I presume, these genera were proposed. I believe the work was never regularly in the market, and can only be procured in an indirect manner. It is a question how far this is a *publication*. I have seen portions of the work in the library of the Linnean Society, but have not met with any indications of the two genera in question. I have, however, received type-specimens through M. Schaufuss, of Dresden.

sions which are common to many Colydiidæ. There is some doubt
as to the tarsi: the anterior has five joints, although the basal one
is only visible from beneath, as shown in the right-hand figure
(Pl. III. fig. 1); but the remainder *appear* to have only four. The
head, from its great breadth, appears to be only very slightly exserted. I have five or six species of this subfamily in my collection,
which I have not yet examined.

Crinc cephalotes. (Pl. III. fig. 1.)

C. ferruginea, nitida; capite prothoraceque vage, elytris seriatim punctatis.
Hab. Ega (Amazons).

Short, depressed, ferruginous, shining, the sides nearly straight, but
gradually becoming narrower from the eye to the last abdominal segment; head and prothorax with large, scattered, shallow punctures,
the latter with a smooth central ovate space, slightly contracted anteriorly, extending from the base to the fore margin, and surrounded by
a line of strong punctures; scutellum nearly triangular; elytra abruptly
rounded at the apex, with about seven rows of oblong punctures on
each; pygidium strongly punctured; legs and body beneath ferruginous; abdomen, except the basal segment, strongly punctured; eyes
dark brown; head large, triangular, deflexed, with a groove extending
from the insertion of the antennæ to the mandibles; epistome very
small, concealing the lip; eyes lateral, prominent; antennæ exposed
at the base, twelve-jointed, the first large, obconic, the second and
third successively smaller, the fourth to the ninth inclusive subequal in
length, but gradually becoming more and more transverse, the tenth
largest of all, and with the gradually diminishing eleventh and twelfth
forming a shortly ovate club; palpi short, stout; labium oblong,
mentum transverse; prothorax transversely quadrate; elytra as broad
as the prothorax at the base; legs short, coxæ subremote, interfemoral
process truncate anteriorly; tibiæ trigonate, the border at the distal end
spinous beneath; tarsi short, the anterior five-jointed, the last as long
as the rest together; abdomen with five segments, the three intermediate very short and equal. Length 1 line.

PHORMESA [Colydiidæ].

Caput insertum, subquadratum, ante oculos dilatatum. *Antennæ* basi
tectæ, clava biarticulata, sulcis antennariis brevibus. *Mentum* quadratum. *Maxillæ* lobis angustatis. *Prothorax* transversus, antice
sinuatus, marginibus dilatatis, crenatis. *Tibiæ* lineares, breviter calcaratæ. *Tarsi* articulis tribus primis brevibus.

It will be seen from these characters that this genus differs but in
few particulars from *Bitoma*; the presence of antennary grooves and
the dilated margin of the prothorax are, however, of too much im-

portance to allow of its being referred to that group. *Bitoma prolata* (*ante*, vol. i. p. 102) belongs to *Phormesa**.

Phormesa lunaris.   (Pl. III. fig. 6.)

P. fusca; prothorace lateribus rotundatis, utrinque bicostato, costis vix elevatis, interiore postice duplicata, exteriore interrupta; elytris luteo bifasciatis.

Hab. New Guinea (Dorey).

Moderately broad, dark brown; head finely and thickly granulose, considerably dilated before the eyes, and hiding the basal joint of the antennæ; mentum quadrate; labium transverse, slightly emarginate; maxillary lobes narrow; prothorax rough, granulated, rather contracted at the base, the disk with two slightly elevated costæ on each side, the interior approximating anteriorly, and forming a short, closed canal towards the head, posteriorly also approximating, then doubling back, and forming a short loop at the base, the exterior costa interrupted in the middle; elytra ovate, wider than the prothorax at its junction, with five crenulated costæ on each, the intervals with a double row of large, deeply impressed punctures, a yellow semilunar band near the middle, and a narrower and straighter one below it; legs pale ferruginous; body beneath dark brown.   Length 1½ line.

On comparison with *Phormesa prolata* it will be seen that, besides the markings on the elytra, the differences will be found chiefly in the prothorax, which in that species is not contracted, except close to the base, and is then a little before the base as broad as the elytra, that the costæ are much more strongly marked, and the outer one especially is entire in its whole length. *Phormesa prolata* is also larger, and proportionally not so broad.

Phormesa inornata.

P. fusca; prothorace lateribus medio subparallelis, basi rotundatis, utrinque bicostato, interiore postice duplicata, exteriore vix elevata; elytris postice obsolete luteo signatis.

Hab. New Guinea (Dorey).

A longer species than the last; the sides of the prothorax less regularly rounded, and broader in proportion to its length, the external costa straighter and nearly entire; the elytra altogether brown, except a very faint spot on each near the base.

* The diagnosis for this will now read thus :—

Phormesa prolata.

P. fusca; prothorace utrinque bicostato, costis fortiter elevatis, interiore postice duplicata; elytris obsolete luteo-maculatis.

Hab. Batchian.

Phormesa demissa.

P. angustior, fusca; capite subreticulato; prothorace lateribus antice rotundatis, dein subparallelis, basi vix constricto, utrinque bicostato, costa interiore postice incurvata; elytris lateribus subparallelis.
Hab. Malabar.

Much narrower than the preceding, brown; head rugosely punctured, with a few irregular and slightly elevated lines, so disposed as to form a kind of network; prothorax broadly margined, the disk with two elevated lines on each side, the interior approximating anteriorly and forming a short canal, strongly incurved at the base, the exterior costa entire; elytra rather broader posteriorly, each with five costæ, the intervals broad and marked with a double row of coarse obscurely defined punctures; body beneath chestnut-brown; legs and antennæ yellowish testaceous. Length 1½ line.

Narrower than the other species of this genus, and easily distinguished from them by the form of the costæ of the prothorax and the reticulated head.

ILLESTUS [Colydiidæ].

Caput quadratum, ante oculos dilatatum. *Oculi* rotundati, prominentes. *Antennæ* articulis duobus primis incrassatis, clava triarticulata. *Palpi labiales* articulo ultimo ovato, obtuso. *Prothorax* subquadratus, irregulariter sulcatus, lateribus marginatis, serrulatis. *Elytra* costata. *Pedes* graciles; *tibiis* anguste trigonatis, calcaratis; *tarsis* brevibus.

Near *Lasconotus* (subfamily *Synchitinæ*), a genus very briefly characterized by Erichson. The eyes, however, are said to be entirely covered by the dilated borders of the head—an unusual structure in this family. Here they are more than usually prominent. In the female of the species described below, the prothorax is more decidedly transverse than in the male.

Dr. Leconte, in the ' Journal of the Academy of Natural Sciences of Philadelphia,' 1859, p. 282, has shortly described a Colydian which he refers to this genus; he observes that it is "at once recognized by its concave head and three-jointed club of the antennæ," but nothing is said in reference to the unusual position of the eyes. It is from Punta de los Reyos in California.

Illestus terrenus. (Pl. III. fig. 4.)

I. fuscus vel rufo-fuscus, opacus; oculis nigris.
Hab. Mexico.

Dark brown or reddish brown, opake; head partially exserted, quadrate, finely granulated; eyes round, prominent, black; antennæ with the basal joint thickened, partially covered at the base, the second also thickened but shorter, the third as long as the first, the remainder to the eighth shorter and more or less transverse, the three last forming an ovate, compact club; maxillary lobes narrow, fringed; mentum subquadrate, rounded in front; labium transverse, narrower behind; terminal joint of the maxillary palpi ovate-triangular, of the labial ovate, obtuse; prothorax somewhat quadrate, but with the sides contracted in the middle, produced at the anterior and slightly emarginate at the posterior angle, the margin rather dilated, especially anteriorly, and serrulate, the disk finely granulated; an elevated line on each side, which are nearly parallel in front, then slightly diverging, after which they approach to form a V-shaped mark, without however becoming connected, each then encloses a lozenge-shaped cavity and terminates at a short distance from the base; outside the line the prothorax is rather concave, with a slight ridge posteriorly; elytra with five strongly marked costæ, the intervals broad, with a double row of coarse punctures; body beneath dark chestnut-brown, reticulate-rugose; legs rather slender; tibiæ gradually thicker towards the extremity and slightly spurred; tarsi short, the first three joints nearly equal. Length 2–3 lines.

NEMATIDIUM [Colydiidæ].

Erichson, Naturg. der Ins. Deutschl. iii. p. 275.

Nematidium mustela. (Pl. III. fig. 10.)

N. ferrugineum; capite antice subdepresso; elytris striato-punctatis. *Hab.* Rio; Para.

Linear, elongate, ferruginous; head finely punctured, moderately convex, somewhat flattened in front, the eyes rather large, black; prothorax half as long as the elytra, finely punctured, the sides slightly incurved; scutellum small, rounded; elytra striate-punctate, the intervals also punctured mostly in an irregular row; body beneath finely punctured; legs luteous testaceous. Length $2\frac{1}{2}$–$3\frac{1}{2}$ lines.

I have no hesitation in considering the insect just described a *Nematidium*, a genus founded on the *Colydium cylindricum*, Fab., and which, but for the expression " *elytris lævissimis*," might have been identical, so far as his short description goes. Whether the *Nematidium costipenne*, J. du Val, really belongs to the genus is, I think, doubtful. I have another *Nematidium* among Mr. Bates's Amazons Colydiidæ *, which differs from the above principally in its more slender form, shorter and more convex head, and elytra more than twice as long as the prothorax. Like *Colydium*, the first abdominal segment is *nearly* as large as the succeeding one. My description is drawn up from the largest of the two specimens now before me, which is from Rio, and belongs to Mr. Fry.

* The Colydiidæ of this collection will form the subject of a distinct paper.

BOTHRIDERES [Colydiidæ].

Erichson, Naturg. der Ins. Deutschl. iii. p. 288.

Bothrideres? rhysodoides. (Pl. III. fig. 11.)

B. ? elongatus, castaneus, nitidus; prothorace lateribus postice angulatis, disco profunde longitudinaliter excavato, basi canaliculato; elytris ovato-oblongis, singulo quinquecostato, costa secunda abbreviata.

Hab. New Guinea (Dorey).

Narrowly elongate, chestnut-brown, shining; head shortly ovate, very convex in front, minutely punctured; eyes large, round, rather prominent; antennæ scarcely longer than the head, the club a little longer than broad, the last joint nearly as large as the preceding one; prothorax rather elongate, the anterior angles produced, the sides rounded, but considerably contracted posteriorly, the disk with a deep ovoid longitudinal impression extending its whole length except a little in front, but which is narrower posteriorly, (there is a very faint trace of a raised central line or space); scutellum punctiform; elytra narrowly ovate, the shoulders a little produced, the base wider than ·the prothorax at its junction, each with five costæ, the first sutural, moderately raised, the second extending to only about a third the length of the elytron, the remainder very strongly elevated, punctation nearly obsolete; body beneath smooth, shining, impunctate; legs moderately long, tibiæ of the anterior and intermediate pairs slightly serrated externally; tarsi about half the length of the tibiæ. Length 3 lines.

Resembles a *Rhysodes* in habit. As the specimen now before me is unique, I must, without an examination of its trophi, satisfy myself with referring it to *Bothrideres.*

Bothrideres? nocturnus. (Pl. III. fig. 12.)

B. ? elongato-ovatus, robustus, castaneus, nitidus; prothorace disco linea parallelogrammum includente impressa; elytris profunde striato-punctatis; antennarum articulo ultimo præcedente majore.

Hab. New Guinea (Dorey).

Elongate-ovate, reddish-chestnut, shining; head considerably exserted, hollowed out between the eyes, thinly punctured, the lip nearly hidden by the clypeus; antennæ not longer than the head, the terminal joint larger in every way than the preceding one; eyes large, very prominent; prothorax scarcely longer than wide, the anterior angles prominent, but not projecting, the sides rounded, much contracted and sinuate at the base, with a deep fovea on each side near the angle, the disk covered with very small distant punctures, and having in its centre a deeply impressed line including a parallelogrammical space; scutellum nearly punctiform; elytra rounded at the sides, the base slightly contracted, but much broader than the prothorax at its junction, striato-punctate, the interstices scarcely raised, except the third one at the base, the first stria much deeper than the others; body beneath chestnut, finely and remotely, the mesosternum and last four abdominal segments coarsely punctured; legs stout; tibiæ short, all strongly spurred, the anterior and intermediate pairs trigonate, dilated and toothed externally; tarsi nearly as long as their corresponding tibiæ. Length 3 lines.

This species is also referred doubtfully to *Bothrideres*, principally on account of the large terminal joint of the club, and the short and unusually trigonate tibiæ; these characters are, however, chiefly ones of degree, and not of plan. An examination of the trophi (which, as the specimen is unique, I have not attempted) might probably afford stronger grounds for its generic separation.

Machlotes [Colydiidæ].

Caput receptum, triangulare, sulcis antennariis. *Antennæ* breves, articulo primo incrassato, libero, clava biarticulata. *Prothorax* sulcatus, postice transversim fissus. *Elytra* ovata, costata. *Pedes* robusti; *protibiis* subtrigonatis, anterioribus spina terminali; *tarsis* brevibus.

A very distinct genus, although, from its widely separated coxæ and large basal segment of the abdomen, allied to *Bothrideres*. The sculpture of the prothorax is, however, peculiar, owing to the presence of a deep transverse cleft posteriorly, dividing, and even dipping below the longitudinal grooves by which the disk is indented. I regret that, having only a single specimen, for which I am indebted to Mr. Bowring, I cannot throw any light on the structure of its mouth, which might perhaps have afforded some clue to its affinities; but if it has no connexion with *Dastarcus*—and even in that case it cannot be a near one—it must remain for the time an isolated genus among the *Bothriderinæ* as they have been defined by Erichson.

Machlotes porcatus.   (Pl. III. fig. 13.)

M. fuscus, opacus; prothorace utrinque tricostato; elytris profunde sulcatopunctatis, interstitiis elevatis.
Hab. Penang.

Dark brown, opake, the antennæ and legs subrufous; head inserted to the eyes in the prothorax, small, and coarsely punctured; antennæ not longer than the breadth of the head, uncovered at their insertion, the basal joint very thick, the remainder more or less transverse, the tenth and eleventh forming a short circular club, of which the last joint is much the smallest; antennary grooves well marked; eyes round; prothorax about half as long again as broad, narrowed behind, truncate and a little gibbous in front, slightly rounded at the sides, the anterior angles prominent,—the disk with three very strong costæ on each side, which are interrupted posteriorly by a deep irregular cleft completely dividing the four central costæ, but less perfectly each of the lateral ones, the fissure moreover in their case extending forward to near the middle of the side, where it forms a deep notch; scutellum punctiform; elytra elongate-ovate, deeply and broadly sulcated, the sulcations pitted with large squarish punctures, the interstices strongly raised and minutely crenate; prosternum coarsely punctured, with a pale curved seta arising from each puncture; meso- and metasternum and abdomen with very large scattered punctures; anterior coxæ widely apart; legs rather robust; the protibiæ with two distinct spines; the tarsi short, with the claw-joint shorter than the three preceding ones. Length 1½ line.

On Plate III. fig. 7, I have represented the trophi of a species of *Dastarcus*, Walker. They are from a specimen given me by Mr. Bowring, who took several individuals at Penang. They differ from *Dastarcus confinis*, Pasc., only in their smaller size, and may safely be referred to that species. The only points I would call attention to, at present, are the *central* insertion of the maxillary palpus (owing, apparently, to a dilatation of the external lobe and its stipes) and the large hook-shaped apex of the internal lobe, not very plainly distinguishable in the figure, owing to the fringe of hairs which borders it, but perfectly distinct in the original.

PETALOPHORA [Colydiidæ].
Westwood, Cabin. of Orient. Entom. p. 85.

Petalophora brevimana. (Pl. II. fig. 9.)

P. nigra, subnitida; prothorace haud canaliculato; elytris singulis sex-costatis; tibiis anticis breviusculis.

Hab. Borneo (Sarawak).

Black, slightly nitid, with the antennæ and palpi reddish pitchy; head rather coarsely punctured, slightly produced below the eyes, with a strongly elevated mesial ridge; epistome not apparent; labrum transverse, subemarginate (not semicircular), fringed with golden-yellow hairs; antennæ fully exposed at their insertion, the club compressed and covered with short hairs; prothorax turgid, subquadrangular, gradually narrower towards the base, the sides straight, the front irre-gular, very obtuse, with a small vertical tooth on each anterior angle; the disk coarsely punctured, not canaliculate, but furnished with a central line, on each side of which at the base are two short diverging ridges; scutellum small, triangular; elytra parallel, gradually rounded at the apex, broader than the prothorax at its base, each with five strongly marked costæ (including the sutural) extending its whole length, and another less marked and shorter at the side, the intervals coarsely punctured; legs robust, the anterior tibiæ very broad and short; posterior coxæ remote, with the first abdominal segment largest. Length 5 lines.

The type of this very rare genus, *Petalophora costata*, is from Java, and differs from the one described above in its canaliculate prothorax, elytra with three costæ only on each, but above all by its having a triarticulate club. Under ordinary circumstances, or if the latter character had been accompanied by any difference in habit, the two could not have been treated as congeneric; as it is, there is such a decided affinity between them, that their separation, except as species, would not be justifiable. *Petalophora*, from the greater size of the basal segment of the abdomen and the widely separated posterior coxæ, must be placed with the *Bothriderinæ* near *Sosylus*, and not with the *Colydiinæ* as has been done in the ' Genera des Coléoptères,' the learned author not having seen it, and Professor Westwood having omitted to give the only two characters by which its position could be ascertained.

Metopiestes [Colydiidæ].

Caput receptum, subverticale. *Antennæ* breves, liberæ, clava biarticulata, compressa, rotundata, sulco antennario laterali. *Prothorax* subovatus, lævis. *Elytra* subparallela, carinata. *Tibiæ* breves, subtrigonatæ, calcaratæ. *Tarsi* elongati, articulo primo majore. *Corpus* cylindricum. (Coxæ posticæ distantes. Abd. segmento primo majore.)

The specimen from which this diagnosis is drawn being unique, I have not been able to examine the parts of the mouth; the genus, however, affords very distinctive peculiarities in its external characters, approximating most nearly to *Petalophora*, but differing in the form of the prothorax, antennæ, &c.

Metopiestes hirtifrons.   (Pl. III. fig. 2.)

M. fusco-castaneus, nitidus; fronte fulvo-tomentosa; antennis rufescentibus.
Hab. New Guinea (Dorey).

Subcylindrical, dark chestnut-brown, shining; head deeply inserted in the prothorax, subvertical, the front densely covered with short fulvous hairs; antennæ 11-jointed, free at their insertion, the basal joint ovate, incrassate, the second longer than the following, pyriform, inserted at the top and side of the first, the rest transverse, the two last forming a round compressed club; antennal groove short, distinct, lateral; eye rather large, ovate; lip transverse; prothorax somewhat ovate, smooth, very convex, rounded in front and at the sides, slightly contracted behind, bisinuate at the base, covered with small, oblong, rather distant punctures, a short semicircular elevated line close to the scutellum; scutellum small, ovate; elytra parallel, rather wider than the base of the prothorax, to which they are closely approximate, each with five very marked elevated lines, the wide excavated grooves between these impunctate, but with a faint trace of another line; body beneath dark chestnut; legs reddish chestnut; femora very robust; tibiæ short, subtrigonal, spurred, the anterior very strongly curved; tarsi elongate, the basal larger than the two following, especially the intermediate and posterior. Length 3½ lines.

Penthelispa [Colydiidæ].
Pascoe, Journ. of Entom. i. p. 111.

Penthelispa Truquii.

P. fusco-castanea, subnitida; prothorace convexo, fortiter punctato, lateribus antice rotundatis, medio paullo constrictis.

Hab. Mexico.

Chestnut-brown, the elytra sometimes with a more reddish tint than the rest, subnitid; head coarsely punctured; antennæ rather stout, the last joint of the club somewhat narrower than the preceding one; prothorax rather longer than broad, the anterior angles produced, the sides rounded anteriorly, but a little constricted in the middle, then again slightly rounded and contracting to the base, the disk convex, without any central depression, and very coarsely punctured; scutellum transversely rounded; elytra broadest nearly at the base, and very slightly rounded at the sides for two-thirds its length, the anterior angle not produced, strongly striato-punctate, the punctures shortly linear; body beneath dark chestnut-brown, shining, very coarsely punctured; legs dark brown. Length 2 lines.

There is a great similarity between the various species of *Penthelispa**, but the prothorax appears to offer good characters by which they may be distinguished. The one described above has that part regularly convex, and free from any impression or any elevated line, and this separates it from the remainder of the few species yet published. I owe my specimens to my kind friend Mr. Fry, who received it together with a vast number of Coleoptera collected in Mexico by the late lamented Signore Truqui, the Italian Minister in that country, after whom I have named it.

Irsaphes [Cucujidæ].
Caput obcordatum, angulis posticis haud productis, collo brevissimo. *Antennæ* moniliformes, articulo primo brevi, tertio paullo longiore.

---

* This name was published in October 1860. Dr. Leconte, in his 'Classification of the Coleoptera of North America,' published at Washington "May 1861—March 1862," proposed the term "*Endectus*" for the North American species.

Mentum transversum, subintegrum. *Palpi* articulo ultimo ovato. *Pro-thorax* subquadratus, lateribus denticulatis. *Tarsi* subdilatati, articulo primo majore. *Corpus* sublatum, planatum.

Allied to *Cucujus* and *Platisus*. The first it strongly resembles in habit, but differs in the normal condition of the tarsi, the ovate terminal joint of the palpi, the head not prolonged behind the eyes, the mentum nearly entire anteriorly, and the *broad, rounded* lobes of its deeply divided labium. From *Platisus* it differs in its robust habit, thicker antennæ, the third joint of which scarcely exceeds the first in length, the narrower tarsi, not dilated at the sides, and the denticulate margins of the prothorax.

Ipsaphes mœrosus.   (Pl. III. fig. 9.)

I. piceo-niger, subnitidus, confertim punctatus; elytris singulis in medio obsolete bicostatis.

Hab. New South Wales.

Pitchy black, subnitid, especially the head and prothorax, finely and very closely punctured; head broadly obcordate, a deep transverse groove behind the eyes, the clypeus descending between the mandibles and hiding the lip; antennæ rather longer than the breadth of the prothorax, moniliform, the basal joint short, incrassated, the second short, the third scarcely longer than the first, the remainder shorter and subequal, the last ovate, pointed; eyes moderate, rounded; maxillary palpi with the terminal joint oblong-ovate, of the labial shortly ovate; maxillary lobes shortly ciliated at the extremity; mentum transverse, not produced anteriorly (the large transverse piece beneath this in the figure is the jugular plate); labium bilobed, the lobes broad, rounded; prothorax subquadrate, broader than long, rounded at the side, with four or five minute, distant teeth, the disk near the anterior angles slightly hollowed out; scutellum transverse, rounded behind; elytra plane, strongly bent down at the sides, each having on its disk two nearly obsolete elevated lines in addition to the more strongly elevated line of the suture; body beneath and legs reddish pitchy, closely punctured. Length 7 lines.

Synœmis [Cucujidæ].

Caput oblongo-subquadratum. *Oculi* prominuli, prothorace distantes. *Antennæ* breves, subclavatæ, articulo basali ovato incrassato. *Maxillæ* lobo interiore uncinato. *Tarsi* articulis tribus primis dilatatis, penultimo minuto. *Corpus* elongatum, parallelum, planatum.

A remarkably elongate and narrow form belonging to the sub-family *Sylvaninæ* as at present constituted, strongly illustrating the impropriety of separating *Sylvanus* from the Cucujidæ, as has been done by M. Jacquelin du Val, and of the danger of coming to conclusions in regard to the limits or characters of natural groups from the examination of the species of a particular region only. M. du Val excludes *Sylvanus* and the cognate genus *Nausibius* from Cucujidæ because their tarsi have not the short basal joint which the remainder of the European members of this family possess; and to this character he attaches an importance of the highest order, so that for him none others are Cucujidæ; but if we look to the well-known genus *Palestes* (and still more to *Ipsaphes* just described), to *Platisus*, or to *Scalidia* and *Ancistria*, where the basal joint far exceeds in size and length those which follow, we shall see at once the utter futility of this character. I think, too, it shows how cautious it is necessary to be before we take what may prove to be a mere technical character for one of real natural import-ance. The division of the Cucujidæ according to the difference of number of the tarsal joints in the two sexes is also objectionable. *Pristoscelis**, which can scarcely be distinguished otherwise from *Pædiacus*, is pentamerous in both, and would therefore be placed by M. du Val with *Monotominæ* †. With regard to *Synæmis*, we must, I think, for the present consider it an isolated genus. The number of these insects, which conceal themselves under bark and in the axillæ of leaves, is probably enormous. They are generally minute, and are not often sought for, and we must therefore expect to find a form turning up now and then whose affinities are uncertain. The posterior tibiæ and tarsi of *Pristoscelis* (accurately described by Mr. Wollaston, but as to the tarsus most inaccurately represented in the figure) are to a certain extent repeated in *Synæmis*; it has also the hooked inner maxillary lobe of that genus. I owe this most interesting form to Mr. Bowring, who took it in considerable abundance at Penang, in the axillæ of the leaves of a species of *Pandanus*.

Synæmis pandani. (Pl. III. fig. 8.)

S. fusco-testaceus, nitidus; prothorace vage punctato; elytris punctato-striatis.
Hab. Penang.

Elongate, very narrow and depressed, chestnut-brown, subnitid; head nearly plane, oblongo-subquadrate, a little broader behind the eyes, sparingly punctured; antennæ remote from the eyes, short, the basal joint thickened, as long as the next two together, the remainder

---

* This name has been preoccupied by Dr. Leconte for a genus of *Dasytinæ*.
† *Monotoma*, according to M. du Val, has 5-jointed tarsi, and he therefore places it with the Cucujidæ.

subtriangular, gradually enlarging to the ninth, which, with the tenth and eleventh, are of equal thickness, the latter a little pointed at the apex; eyes prominent; mentum transverse, narrowed in front, its anterior angles produced; labium slightly emarginate; maxillary lobes narrow, nearly equal in size, fringed with long hairs, the inner lobe with a strong hook at its external angle; palpi rather short, the terminal joint of the maxillary subcylindric, of the labial ovate; mandibles bifid at the apex, with a slender tooth internally; prothorax twice as long as the head, sparingly punctured, a small process at the anterior angle, posteriorly a little contracted, and at the base a curved impressed line; scutellum broadly triangular, the sides rounded; elytra about twice as long as the prothorax, punctate-striate, slightly concave between the suture and the external border, where they bend down almost at a right angle; coxæ not approximate; femora long, robust; tibiæ short, slightly curved, subtrigonate, the posterior near the extremity finely toothed at its inner edge; tarsi very short, the three basal joints dilated, the fourth minute, the claw-joint small, not longer than either of the three basal; body beneath dark brown, finely punctured. Length 3 lines.

The insect is much narrower than I have represented in the figure.

ACHTHOSUS [Tenebrionidæ].

Caput exsertum, clypeo producto. *Antennæ* subclavatæ, articulis 5–7 ultimis perfoliatis, transversis. *Maxillæ* lobo interiore hamato. *Tibiæ anticæ* trigonatæ, extrorsum dentatæ. *Corpus* subcylindricum.

This genus differs in a few points only from *Antimachus*, some species of which it closely resembles, except that it is more cylindrical, but from which it will be at once distinguished by the strongly serrated external margin of the fore-tibiæ. There are also remarkable differences in the mentum and labium of the species described below, and in the same parts of a species of *Antimachus* (probably *A. furcifer*, Gistl) which I examined for the purpose of comparison. But two other species, which I refer also to *Achthosus*, appear to have the more or less subcordate mentum of *Antimachus*, and therefore I have not referred to this organ in the characters of the genus. So far as my limited experience goes, it appears to me that the parts of the mouth are subject to the same variations as other organs, and, except certain differences of plan, which, however, are rather characteristic of higher groups than genera, the variation in form or outline of these organs is generally only one of degree. I believe that they are supposed to be more constant in their characters because they are seldom examined, and that one species is, as a matter of course, taken as the type of the rest. For this reason

I have generally avoided entering into details of these organs in the
generic characters, reserving them for the species which alone has
been examined. If I have correctly recognized the sexes, there
appears to be little difference between them, at least in the species
described below. This Tenebrionid is not rare in collections: Professor Westwood informs me that it stands in the Oxford Museum as
Dendroblaps Westwoodii (Macleay). This name has not been published, I believe; and as there is a *Dendroblax* among the Lucanidæ,
I have retained the generic name under which it has always stood
in my cabinet.

Achthosus Westwoodii. (Pl. II. fig. 7.)

A. niger, nitidus; clypeo recurvato; prothorace antice excavato, margine
supra trisinuato.

Hab. Australia.

Subcylindrical, deep black, shining; head a little dilated anteriorly,
narrowed behind the eyes, where it forms a thick neck, the front
slightly concave and somewhat finely punctured, the clypeus produced and slightly recurved; epistome very distinct, subquadrate, the
lip obsolete; antennæ with the five or six last joints perfoliate, transverse, and considerably broader than the others; mentum stout and
irregular, but with six nearly equal sides; labrum somewhat cordate,
its palpi inserted in a cavity which is hollowed out on each side at its
base; last joint of the maxillary palpi shortly triangular, of the labial
obliquely ovate; prothorax slightly broader than long, strongly excavated anteriorly, and this part only thickly punctured, the border of the
excavation posteriorly strongly marked and having a trisinuate outline; scutellum cordate-triangular; elytra parallel, coarsely punctate-
striate, the intervals broad and nearly impunctate; body beneath black,
shining; antennæ and legs chestnut; anterior and intermediate tibiæ
strongly serrated externally, the posterior only very slightly so, all
terminated by two or three stout spines; tarsi narrow, the claw-joint
as long as the rest together. Length 10 lines.

Strongylium [Tenebrionidæ].
Kirby, Trans. Linn. Soc. xii. p. 417.

Strongylium Macleayi.

S. nigro-chalybeatum, nitidum; prothorace transverso, antice rotundato,
basi angustiore; scutello nigro-cupreo; elytris subelongatis, seriato-
punctatis, lateribus parallelis.

Hab. New South Wales.

Dark chalybeate blue, shining; head finely punctured; eyes nearly
contiguous above; epistome and lip bordered with testaceous; antennæ about half the length of the elytra, the third joint much longer
than the first and second together, the fourth and fifth gradually

shorter; prothorax finely punctured, much broader than long, considerably rounded at the anterior angles, the sides gradually but slightly narrowing posteriorly, a shallow fovea on each side in front; scutellum dark copper-brown; elytra seriate-punctate, the punctures coarse, rather elongate, the sides parallel for about two-thirds of their length, then slightly rounded and gradually tapering to the apex; body beneath and legs dark brown or black, with a tinge of reddish, especially on the femora; posterior tarsi with the basal joint longer than the rest together. Length 6 lines.

There are very few species of this genus described in comparison to those in collections; and none, I believe, from Australia. I do not know anything to which the one here described can be assimilated, except one from Mysol, which, however, has only a certain similarity of outline.

CAMPOLENE [Tenebrionidæ].

Caput subexsertum, antice dilatatum, postice paullo constrictum. *Oculi* parvi, emarginati. *Antennæ* breves, claviformes. *Tibiæ* curvatæ, muticæ. *Prosternum* antice constrictum, postice subhorizontale, incurvato-productum. *Mesosternum* declinatum, antice triangulari excavatum.

These characters are intended to be contrasted with those of *Chariotheca* and *Titæna*, between which, I believe, this genus should be placed. The unarmed tibiæ, and the partially horizontal and then incurved posterior portion of the prosternum, terminating in a short triangular process very imperfectly received in the corresponding notch of the mesosternum, will distinguish it from the former: while in *Titæna* the anterior portion of the prosternum is so contracted that it forms a mere line in front of the two cotyloid cavities, so that the head in repose rests on the coxæ, this part has the normal form in *Campolene*. There are also other differential characters which it is not necessary to mention now. In habit *Campolene* resembles *Helops*.

Campolene nitida.   (Pl. II. fig. 4.)

C. elongato-ovata, nigra, nitida; prothorace subtiliter, elytris seriatim punctatis; pedibus rufo-ferrugineis.
Hab. New South Wales.

Elongate-ovate, black, shining; head finely punctured, slightly contracted behind the eyes, expanded and a little concave anteriorly, the lip nearly hidden beneath the clypeus; antennæ shorter than the prothorax, the third and fourth joints longest, the rest becoming gradually shorter, broader, and more compressed, the last largest and nearly circular; eyes small, lateral, emarginate in front; terminal joint of

the maxillary palpi securiform, of the labial narrowly triangular; prothorax finely punctured, convex, slightly transverse, rounded anteriorly
and laterally, and narrowly margined; scutellum small, triangular;
elytra coarsely seriate-punctate, scarcely broader at the base than the
prothorax, the sides gradually rounded to the apex; body beneath with
the sterna dull reddish ferruginous, the abdomen glossy black; prosternum subhorizontal posteriorly, incurved, ending in a short thick
process which is only partially received in the shallow corresponding
notch of the mesosternum: intercoxal process rather broadly triangular;
legs reddish ferruginous, rather slender; tibiæ strongly curved, and unarmed; tarsi narrow, hairy beneath, the basal joint slightly elongate,
the last shorter than the preceding united.   Length 4 lines.

Apellatus [Cistelidæ].

Caput antice elongatum; *oculis* magnis, reniformibus. *Antennæ* breves,
articulo primo vix incrassato, tertio ad septimum subæqualibus, haud
nodosis. *Tibiæ* breves, curvatæ. *Prosternum* compressum, elevatum.

The genera of Cistelidæ do not appear to be distinguished from
each other by any very trenchant characters. This genus is perhaps
scarcely an exception, although in colour it differs essentially from
*Æthyssius** and *Tanychilus*, genera to which, on account of their
long muzzle, this is the most nearly allied: from these, and especially
from the latter, it is separated by its shorter antennæ, with the
basal joint scarcely thickened, the nearly equal length of the third
to the seventh inclusive, their subcylindrical form (not nodose at the
end), the shorter and curved tibiæ, the larger and more reniform
eyes, and the narrow prosternum.   I only know the males.

Apellatus lateralis.   (Pl. II. fig. 1.)

A. flavo-testaceus, glaber, subnitidus; oculis vittaque elytrorum nigris.
Hab. New South Wales.

Fulvo-testaceous, smooth, subnitid, a stripe from the shoulder
gradually widening behind, and at the apex nearly approaching the
suture, and eyes black; head narrow, prolonged beyond the eyes, and
rounded immediately behind them; antennæ about half as long as the

---

* *Æthyssus*, proposed for *Atractus*, Lacord. (Macleay, Dejean), which name
has been in common use since 1832 for a genus of Hemiptera. The name of
another Heteromerous genus (*Trigonotarsus*, Hope) having been preoccupied by
Guérin for a genus of Curculionidæ, I have now to propose " *Sobas*," which I
have used in a MS. list of the Australian Heteromera that I have in hand.   I
have also in the same list adopted as a genus the division distinguished by two
spurs to the anterior tibiæ, which M. Lacordaire has made in *Nacerdes*, and have
named it " *Sessinia*."

body in the male (probably shorter in the female), the basal joint
scarcely thickened, the second short, the third to the seventh of nearly
equal length, subcylindrical, not nodose at the ends, and the remainder
a little shorter and somewhat compressed (except the last, which is
pointed); palpi brownish, the terminal joint of the maxillary securiform,
of the labial shortly triangular; eyes large, reniform; prothorax rather
longer than broad, rounded at the sides, truncate and considerably
contracted in front, finely punctured, two foveæ at the base and an
intermediate depression, posterior angle acute; scutellum triangular;
elytra striate-punctate, much wider than the prothorax, ovate-elongate; body beneath fulvous, pubescent; prosternum narrow, elevated;
mesosternum V-shaped; legs short; tibiæ slightly curved, terminating
in two short spines; the two penultimate of the anterior and intermediate and the penultimate only of the posterior tarsi lamellate.
Length 4 lines.

Diacalla [Lagriidæ].

Caput trigonatum, ad angulum posticum productum.   *Oculi* parvi, rotundati.   *Labium* quadratum, membranaceum.   *Palpi labiales* articulo
ultimo subcylindrico.   *Prothorax* late ovatus, antice constrictus.   *Tibiæ*
bicalcaratæ.

These characters (and there are also others) are in complete oppo-
sition to *Lagria*, with which genus only—if, perhaps, we except
Euomma—in the four which have hitherto composed this family, is
it to be assimilated.   In other respects it agrees perfectly with the
characters of the Lagriidæ as laid down by M. Lacordaire, except
that the eyes are entire, and the labium is so thin and transparent
as to be rather membranous than corneous*.   The habit of the
species described below is more that of a *Titæna* than a *Lagria*.

Diacalla comata.   (Pl. II. fig. 6.)

D. rufo-fusca, subnitida, hirsuta, fortiter et confertim punctata; abdomine infra subrufescente.
Hab. Queensland.

Dark reddish brown, subnitid, closely and very coarsely punctured,
with short erect greyish and black hairs, mostly arising from the
punctures, covering the whole upper surface; head inclined, trigonal,
enlarged behind the eyes, then suddenly contracted into a thick neck;
eyes small, round; antennæ short, the two basal joints slightly thickened, the remainder to the tenth gradually diminishing in length but
increasing in thickness, the eleventh more slender and as long as the
two preceding together; internal maxillary lobe narrow, longer than

---

* Fabricius, however, says "labium membranaceum." (*Ent. Syst.* i. pars ii.
p. 78.)

the outer, both densely ciliated, their palpi long, with the last joint securiform; labium thin, quadrate, fringed anteriorly, its palpi sub-filiform, rather elongate, arising from near the centre of the labium; mentum subtransverse, rounded at the sides, peduncle of the jugular plate as broad as the labium; prothorax broadly ovate, constricted in front, so as to form a sort of collar; scutellum triangular; elytra much broader than the prothorax, gradually tapering behind, rounded at the apex; legs rather short, tibiæ terminated by two spines, basal joint of the anterior tarsi short, the intermediate and posterior gradually longer; body beneath slightly hairy, the abdomen with a reddish tinge. Length 5 lines.

The above description is from a female. A male which I believe belongs to this species is smaller, more hairy, the terminal joint of the antennæ much longer, and the abdomen without the reddish tinge.

Goëtymes [Cantharidæ].

Caput magnum, fronte convexa; *oculis* reniformibus. *Antennæ* breves, frontales, articulo primo subtrigono, incurvato, in sulco infra oculos recepto, secundo tertioque brevibus, reliquis flabellatis. *Tibiæ* uni-calcaratæ. *Tarsi* breves, unguiculis simplicibus.

The nearest ally of this genus is *Sitarida*, White, from which, *inter alia*, it differs, as it does from every other of the family, in its flabellate antennæ, which resemble *Evaniocera* in the nearly allied group of Rhipophoridæ. The difference between the antennæ of the two genera, however, requires to be more clearly contrasted. In both they are 11-jointed; but in *Sitarida* the first four are simple, while each of the remaining seven throws out laterally and at the base a short square lamina—this portion of the antenna being, in fact, pectinate. In *Goëtymes*, the first three joints only are simple, the remainder being drawn out into long laminæ, closely applied to each other at the base, and forming a compact mass when at rest. For the protection of this delicate part in repose, there is a groove beneath the eye, which receives the basal joint, and thus allows the whole antenna to be kept well under the head and breast; and this purpose is facilitated by the antenna not arising in the space formed by the emargination of the eye (which, I believe, is almost invariably the case whenever that organ is reniform or emarginate, and which is apparently so constructed for the express purpose), but below this space, and in front of the inferior portion of the eye. It may be added that the emargination above mentioned is occupied by a short, obtuse process, a simple development of the front.

Goëtymes flavicornis.   (Pl. II. fig. 5.)

G. pallide fulvescens ; mandibulis, prothorace, sternis femoribusque nigris ; antennis flavescentibus.

Hab. Australia (Port Stephens).

Pale brownish fulvous, more or less clothed with short erect hairs ; mandibles, prothorax, breast, and thighs black or brownish black, abdomen and antennæ pale yellow ; head convex and rounded in front, covered with minute vermicular folds ; epistome and lip trigonal ; mandibles thick, bifid at the end, coarsely punctured at the base ; palpi robust, the labial much smaller than the maxillary, the last joint in both ovate ; prothorax subtrigonate, the sides slightly rounded ; scutellum triangular, the apex prolonged into a short quadrate process ; elytra very short, spatulate ; legs robust ; all the coxæ contiguous ; femora and tibiæ ciliated beneath, the latter with a single spur ; tarsi short, the claws simple ; abdomen corneous, not contracting when dry. Length 10 lines.

The specimen described is in the British Museum. The hind tarsi are unfortunately wanting ; in the figure they are assumed to resemble those of *Sitarida Hopei*. Port Stephen or Stephens is about two degrees N. of Sydney.

Cyphagogus [Brenthidæ].
Parry, Trans. Ent. Soc. v. p. 182.

Cyphagogus advena.

C. rufo-testaceus, nitidus ; capite lato, breviusculo, apice emarginato : elytris striatis, striis modice punctatis.

Hab. Natal.

Reddish testaceous, shining ; head as broad as the prothorax, but considerably shorter, finely and sparsely punctured, widely emarginate at the apex, which is bilobed on each side ; eyes round, black ; antennæ scarcely longer than the head ; prothorax narrow, compressed anteriorly, with a few minute, scattered punctures ; no visible scutellum ; elytra as broad as the prothorax, deeply striated, the striæ with shallow, rather distant punctures ; body beneath more coarsely punctured ; legs with the posterior tibiæ not longer than the basal joint of the tarsi of the same pair. Length 3 lines.

This adds one more to the list of remarkable genera common to the Indian Islands and to Natal, yet still sufficiently distinct to form another category in this curious and very strongly marked genus. That is to say, that in its shorter head and thicker rostrum it recedes from *Cyphagogus* and approaches *Zemioses*, which, however, has legs of the more normal character.

MACROTOMA [Prionidæ].

Serville, Ann. de Soc. Ent. de Fr. i. p. 137.

Macrotoma servilis.

M. fusco-castanea, subnitida : prothorace transverso, lateribus submuticis, antice tridentatis, postice unispinosis ; scutello postice rotundato ; elytris connexo-punctatis, haud vermiculatis ; abdomine glabrato, polito. *Hab.* Australia (Melbourne).

Dark chestnut-brown, subnitid ; head coarsely punctured ; antennæ longer than half the length of the body, all the joints more or less punctured, the third nearly as long as the two next together ; prothorax shortly transverse, irregularly and coarsely punctured, the middle portion of its sides straight, but gradually diverging to the base, nearly meeting, anteriorly with three teeth, posteriorly with a spine, at the base of which are two or three short teeth ; scutellum rounded posteriorly ; elytra much broader than the base of the prothorax, the sides slightly rounded, closely punctured, the punctures becoming coarser and more or less connected, although never vermiculate, as they approach the suture and base, this part also being darker or somewhat pitchy ; abdomen and legs pale chestnut, highly polished ; metasternum thinly pilose, prosternum coarsely punctured. Length 18 lines.

The only *described* Australian Prionid that approaches this is *Hermerius impar* of Newman, which, *inter alia,* differs in its hairy prothorax and the thick mass of woolly pubescence which clothes the abdomen. I have not adopted the genus, however, from the impossibility of seeing how it is to be separated from some forms of *Macrotoma.* There are several undescribed species from Australia, differing from each other in a not very tangible manner, but mostly having the sides of the prothorax more denticulate. I fear, however, that the amount of denticulation is very often, in this family, a character varying according to the individual. In the specimen just described, the two posterior teeth of the anterior angle of the prothorax are distinctly bifid on the right side, but are entire on the left. So in Mr. Newman's genus *Cnemoplites**, the teeth on the protibiæ, in a specimen of an undescribed species in the British Museum, are five on one side, and three on the other : in an allied species the intermediate tibiæ are also toothed, and in my *Mallodon figuratum* all the tibiæ. The Prionidæ, as they are constituted at present, appear to be a very unsatisfactory family, containing several anomalous genera, and others which are extremely difficult to limit.

* Mr. Newman describes *Cnemoplites* thus : " *Protibiis excurvatis, extus spinosis* " (Entom. p. 351) ; and, in addition to *C. edulis* (unknown to me), refers to it *Prionus spinicollis,* Macleay, which has *all* the tibiæ spined, and which I cannot separate from *Macrotoma.* It is, in fact, very near my *Macrotoma gemella.*

One of these, *Neostenus* (Trans. Ent. Soc. ser. 2, iv. p. 91), on account of the position of the anterior coxæ, I am disposed to place with the Cerambycidæ, perhaps not far from *Bimia*. This last, also, is a very isolated genus.

OBRIDA [Cerambycidæ].

White, Stokes's Voyage, App. i. p. 510.

Obrida comata.

O. nigro-chalybeata, sparse griseo-pubescens, hirsuta; elytris singulis macula magna mediana flava.

Hab. Queensland.

Very dark steel-blue, lightly covered with a pale greyish pubescence, with scattered, erect, stiffish hairs interspersed; head and prothorax roughly and closely punctured, the anterior and posterior margins of the latter of nearly equal breadth; scutellum triangular, covered with long silky hairs; elytra short, broader than the prothorax, the sides parallel, each furnished with two not very prominent costæ, and in the middle a large transverse yellow spot not attaining the margin or the suture; body beneath shining steel-blue, sparingly punctured with a few scattered hairs; legs more or less hairy, the femora shining steel-blue, base of the posterior testaceous; tarsi rufous brown; antennæ entirely black, about two-thirds the length of the body. Length 4 lines.

Perfectly homogeneous with *Obrida fascialis*, but broader and more robust, with the antennæ and legs entirely black (except the base of the posterior femora), and the broad orange band on the elytra of the former replaced by two pale-yellow patches; it is also more pubescent, furnished with long scattered hairs.

PYRESTES [Cerambycidæ].

Pascoe, Trans. Ent. Soc. ser. 2, iv. p. 96.

Pyrestes cardinalis.

P. ruber, nitidus; scutello, pedibus corporeque infra nigris.

Hab. Hong Kong.

Dark red, brighter on the elytra, shining, with a pubescence consisting of a few short black hairs, but more numerous on the prothorax; head dark brownish-red, thickly punctured; antennæ dark brown, the basal joints coral-red, except at their extremities; eyes black; prothorax about half as long again as broad, rugosely punctured, the punctures large and irregular; scutellum narrowly triangular, black; elytra dark blood-red, coarsely and deeply punctured at the base, but gradually more scattered and shallower towards the apex; legs black, covered with short stiff fulvous hairs; body beneath black, shining, moderately punctured, slightly hairy. Length 7 lines.

In 1857 I briefly characterized this genus, at the same time describing three species, all Asiatic. I do not see that I can add anything really essential to those characters now. The genus is a very natural one, and is allied to *Erythrus*, but with an ovate-elongate or almost subcylindrical prothorax; elytra slightly contracted in the middle, much more convex, and with a broad emargination externally near the shoulder. The palpi also are longer and more unequal. The antennæ vary in length, but are longest in the males, although scarcely so long as the body. The pro- and mesosterna are simple. Professor Westwood has given an excellent figure of *Pyrestes eximius* in the work above quoted (pl. 22. fig. 3).

ERYTHRUS [Cerambycidæ].
White, Cat. Col. Ins. Brit. Mus. Longicornia, p. 142.

Erythrus congruus.

E. niger; prothorace elytrisque coccineis, illo nigro sex-maculato et medio breviter carinato.

Hab. Hong Kong.

Slightly depressed, irregularly and closely punctured, black; prothorax and elytra bright scarlet, the former nearly equal in length and breadth, with six black spots, four on the disk and one on each side, the middle with a short elevated line; scutellum transverse; elytra moderately long, an elevated carina running from each shoulder to near the apex, which is rounded with its edges minutely serrated; body beneath entirely black, very closely and irregularly punctured; legs black, tarsi of the intermediate pair longer than their tibiæ. Length 9 lines.

From *Saperda*? *bicolor*, Westw., this insect differs in being *entirely* black beneath, in its six-spotted prothorax with a short elevated line in its middle, in the more decidedly elevated and longitudinal carina which occurs on each elytron, and in the general vitreous sort of transparency which in certain lights and under a strong lens glistens over its surface, especially on the elevated lines of the prothorax and elytra. It will serve to show the uncertainty of characters generally thought to be of generic value among the Longicorn families that, notwithstanding the close affinity of these two *Erythri*, amounting at the first glance almost to identity, the one, *E. bicolor*, has the epistome very distinct, while in the other it is apparently wanting. *Erythrus Fortunei*, White (the only other *Erythrus* having the head black), is a narrower and smaller species, with a longer prothorax and darker colour.

Erythrus? Bowringii.

E. ? angustatus, rubro-sericeus ; prothorace ovato, medio carinato ; elytris
elongatis, apice truncatis ; corpore infra nigro, griseo-pubescente.
Hab. Hong Kong.

Narrow and elongate, brick-red, covered with a fine silky pubescence ; head roughly punctured, the muzzle rather short ; antennæ
black, longer than the body in the male, about three-quarters of its
length in the female, the serration beginning with the fourth joint ;
prothorax ovate, a long linear carina in the middle, two black spots
anteriorly on the disk, marking the nearly obsolete tubercles ; scutellum
triangular ; elytra elongate, scarcely wider than the prothorax, the sides
incurved and expanding very slightly posteriorly, the apex truncate, a
broadly elevated line extending from the shoulder to near the apex ;
body beneath black, closely covered with a short greyish-white pubescence ; legs black, slightly pubescent, femora of the intermediate pair
produced beneath, and fringed at the deepest part of the border with
short stiff hairs.   Length (♂) 9, (♀) 11 lines.

This species rather breaks in upon the homogeneity of *Erythrus*,
but I scarcely see sufficient characters to warrant its separation as a
distinct genus.   The narrow form, the ovate prothorax, and the
serrated portion of the antennæ beginning at the fourth joint instead
of the fifth, seem to be the most distinctive points.   The muzzle is
also somewhat shorter and the palpi longer, but I think it would be
difficult to formulate a satisfactory diagnosis on these.   The peculiarity of the intermediate femora is less marked in the female.   I
am indebted for this and the two preceding species, and indeed for
many others, to John Bowring, Esq.

POLYZONUS [Cerambycidæ].

Laporte de Castelnau, Hist. Nat. des Ins. Coléop. ii. p. 438.

Polyzonus pubicollis.

P. obscure niger ; prothorace subcylindrico, aureo-pubescente ; elytris
luteis, fasciis tribus, postica subapicali suturam non attingente, nigris.
Hab. Natal.

Dull black ; head coarsely punctured, with a few scattered yellowish
hairs ; epistome very short, lip narrow, bordered with stiff yellowish
hairs ; prothorax short, subcylindrical, slightly narrowed behind, closely
and coarsely punctured, and covered with a golden-yellow pile ; scutellum acutely triangular ; elytra very finely and closely punctured,
sparsely pubescent, luteous yellow, a black band near the base, a second
at the middle, and a third towards the apex, but which does not attain
to the suture ; body beneath black, more or less covered with a silvery-
grey pile, the last abdominal segment extending beyond the elytra ;
legs black, more or less pubescent ; femora scarcely clavate, the posterior

not at all; tibiæ short, the distal extremity of the posterior scarcely reaching to the end of the abdomen; antennæ black, the basal joints with a slight pubescence. Length 9 lines.

Of the two species of *Promeces* mentioned by Serville, one, the *Saperda clavicornis* of Fabricius, is a *Polyzonus*. The error is the more remarkable, as he has perfectly well distinguished *Promeces* by the setaceous, twelve-jointed antennæ of the males. *Polyzonus clavicornis*, a common Cape insect, on the contrary, has the antennæ claviform and eleven-jointed in both sexes. The Comte de Castelnau has failed to notice any peculiarity in the antennæ either of *Promeces* or *Polyzonus*, and is apparently ignorant of the females of the former, since he ascribes filiform antennæ to both sexes, the fact being that they are *setaceous*, not *filiform*, in the males and clavicorn in the females. With regard to *Polyzonus*, the species described above is remarkable for its subcylindrical prothorax rather closely covered with a short decumbent pile, and is distinguished from all others of the genus known to me by the yellow apex of the elytra.

Polyzonus scalaris (Dej.).

P. angustus, chalybeatus; prothorace breviter subovato, rugoso-punctato; elytris luteis, fasciis tribus latis chalybeatis.
Hab. Cape of Good Hope.

Narrow, dark steel-blue; head coarsely punctured, epistome very short, lip large, broader anteriorly, scarcely emarginate, eyes black; prothorax shortly subovate, very roughly punctured, scarcely pubescent; scutellum narrowly triangular; elytra strongly and closely punctured, luteous yellow, with three broad dark chalybeate bands, the first towards the base, the second in the middle, the third apical; body beneath steel-blue, with a silvery-grey pubescence; legs steel-blue, femora of the anterior and intermediate pairs only moderately clavate; antennæ very dark steel-blue. Length 7 lines.

In the disposition of the bands on the elytra this species comes nearest *Polyzonus Mellyi*, White, but is smaller, narrower, with a more ovate prothorax, which is scarcely or not at all pubescent, and with very much broader bands on the elytra. I believe it to have been hitherto unpublished.

PROMECES [Cerambycidæ].
Serville, Ann. de Soc. Ent. de Fr. iii. p. 27.

Promeces viridis (Dej.).

P. viridi-cæruleus, corrugatus; prothorace brevi, lateribus irregulariter rotundatis; femoribus posticis subclavatis.
Hab. Natal.

Dark greenish blue, the whole upper surface finely corrugated ; head coarsely punctured in front, epistome dark brown, shining, lip rounded, covered with greyish hairs, eyes black ; prothorax scarcely longer than broad, irregularly rounded at the sides ; scutellum triangular, very concave ; elytra nearly parallel, without raised lines ; body beneath shining chalybeate blue, sparsely pubescent ; femora blue, the posterior very slightly clavate ; tibiæ and tarsi blue, covered with short stiff hairs, claws reddish testaceous ; antennæ blue, the basal joint coarsely punctured, the last four joints in the female very short and thick.   Length 5 lines.

This long-known species has not, so far as I know, been hitherto described.   It may be at once distinguished from its congeners by its short and corrugated prothorax ; but, like the others, its colour is more decidedly blue than green.

Apodasya [Lamiidæ].

Caput parvum, verticale ; *oculis* emarginatis.   *Antennæ* pilosæ, articulo basali subcylindrico, tertio longissimo, cæteris brevissimis.   *Prothorax* gibbosus, subquadratus, lateraliter spinosus.   *Elytra* parallela.   *Tarsi* breves.   *Pro-* et *mesosternum* simplicia, acetabula antica angulata.   *Corpus* subelongatum.

Chætosoma pilosum of Dejean's Catalogue is the type of this genus, but as the generic name has been used for one of the Cucujidæ, it is necessary to substitute another.   In the above work it was placed between *Desmiphora* and *Cloniocerus*, but it appears to me to be more nearly related to *Hebestola*.   It is not mentioned by M. James Thomson in his ' Essai,' &c. ; indeed it seems to be a very scarce insect, only to be seen in a few old collections.   My specimen is from the collection of Mr. Waterhouse.

Apodasya pilosa.

A. ferruginea, grisescente-pubescens, pilis longis albis nigrisque tecta ; prothorace disco nigro ; antennis pedibusque infuscatis.
Hab. South Africa.

Ferruginous, covered with a very fine greyish pubescence, and with long erect white hairs mingled with black ; head rather small ; epistome and lip very distinct, the latter rounded anteriorly ; palpi pointed ; eyes deeply emarginate ; antennæ very hairy, arising from two diverging tubercles, shorter than the body, the basal joint subcylindrical, the third as long as the rest together, a dense fascicle of black hairs enveloping the fourth joint and apex of the third ; prothorax short, irregularly gibbous, a strong tooth on each side posteriorly, the disk with a large black spot ; scutellum very small, black ; elytra parallel, elongate, broader than the prothorax, very coarsely punctured ; body beneath yellowish ferruginous, the sides of the metathorax and base of the abdomen brown : legs brownish.   Length 5 lines.

Aproïda [Hispidæ].

Caput pone oculos subelongatum; fronte brevi, verticali; clypeo bilobato, labrum occultante. *Oculi* ovati. *Palpi maxillares* articulis ultimis duobus globosis. *Palpi labiales* articulis ultimis oblongo-ovatis. *Mentum* quadratum. *Antennæ* filiformes, super tuberculis inter oculos insertæ, articulis duobus basalibus brevibus, primo incrassato, cæteris brevioribus, ultimo paullo longiore apice appendiculato. *Prothorax* quadrilateralis, postice latior. *Elytra* deplanata, subtrigona, thorace latiora, apice caudata. *Pedes* breves; *femoribus anticis* incrassatis, dentatis; *tibiis* ejusdem curvatis, introrsum bispinosis. *Corpus* subplanatum.

This is probably the most remarkable genus of the Hispidæ, wholly distinct in habit from any other known species, although most nearly related to *Eurispa*. The prolongation of the head behind the eyes, the size and figure of the anterior femora, the two formidable teeth on the protibiæ (as is also the case in some Cephalodontæ), and trigonate outline of the elytra terminating in two thick spines, combine to produce a form that, taken in conjunction with the congeners of its own family, renders it one of the most striking of the Australian Coleoptera. The parts of the mouth can only be described as they are seen *in situ*, and these are the more difficult to distinguish as they are placed in a deep cavity formed by the mandibles in front, and by the jugular plate, bent down at a right angle, behind; it may be also noticed that the angle itself is bordered by an elevated, narrow ridge. I am indebted to Mr. Baly, who is so well known for his Monograph of this family and for his knowledge of the Phytophagous groups in general, for his assistance in this examination; he is satisfied of the existence of a small square mentum which is attached to the anterior edge of the reflected portion of the jugular plate, and that the last two joints of the maxillary palpi are together of a globose form, and those of the labial oblong-ovate.

Aproïda Balyi. (Pl. II. fig. 8.)

A. flavescens, vitta fusco-purpurea ab oculis ad apicem elytrorum ornata; antennis fusco-purpureis, articulis duobus ultimis albis.
Hab. Queensland.

Fulvous, on the elytra inclining to lemon-yellow, a dark-purple line extending from the eye to the apex of the latter; head coarsely punctured, elongate behind, the front vertical, with a tubercle before each eye, bearing the antennæ; eyes ovate, prominent; antennæ about half the length of the body, dark chestnut-brown, the last two joints pale straw-yellow, the basal joint short, incrassate, the second about the same length as the first, the remainder longer, cylindrical, the last terminated by a small hooked appendage; prothorax quadrilateral, broader behind, bulging at the sides, the disk concave near the base and very coarsely punctured; scutellum subtriangular; elytra trigonate, depressed, covered with large rough punctures, broadest at the shoulders, where they considerably exceed the prothorax, gradually contracting towards the apex, and terminating on each side in a stout diverging spine, which is considerably strengthened by a short raised line or rib which connects it with the rest of the elytron; body beneath saffron-yellow, nearly impunctate; mouth, mandibles, and palpi dark brown; the intermediate and posterior legs short, the anterior much longer; femora clavate, with a large obtuse tooth beneath, except the posterior; anterior tibiæ slender, curved, dilated at the apex, with two acute teeth on the inner side.   Length 6 lines.

EXPLANATION OF THE PLATES.

PLATE II.

Fig. 1. *Apellatus lateralis.*
  „ 2. *Diatelium Wallacei.*
  „ 3. *Clidicus formicarius.*
  „ 4. *Campolene nitida.*
  „ 5. *Goëtymes flavicornis.*
  „ 6. *Diacalla comata.*
  „ 7. *Achthosus Westwoodii.*
  „ 8. *Aproïda Balyi.*
  „ 9. *Petalophora brevimana.*

PLATE III.

Fig. 1. *Crine cephalotes.*

Fig. 2. *Metopiestes hirtifrons.*
  „ 3. *Temesia Batesii* *.
  „ 4. *Illestus terrenus.*
  „ 5. *Narcisa decidua.*
  „ 6. *Phormesa lunaris.*
  „ 7. *Dastarcus confinis* (trophi).
  „ 8. *Synarmis pandani.*
  „ 9. *Ipsaphes mœrosus.*
  „ 10. *Nematidium mustela.*
  „ 11. *Bothrideres? rhysodoides.*
  „ 12. *Bothrideres? nocturnus.*
  „ 13. *Machlotes porcatus.*

* The description of this insect will be given in a future Part.

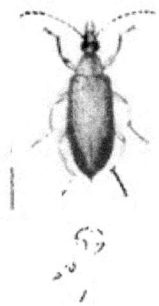

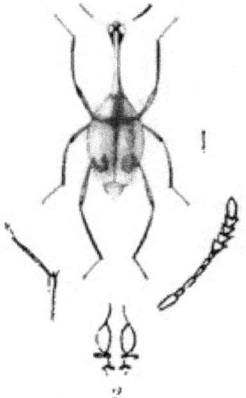

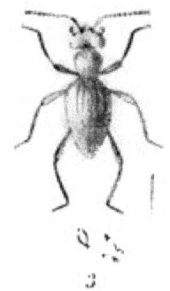

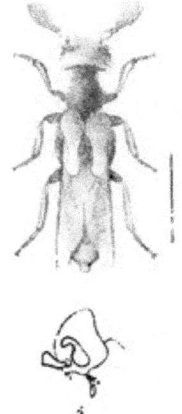

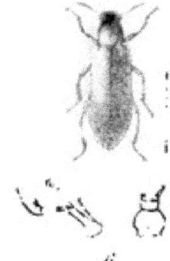

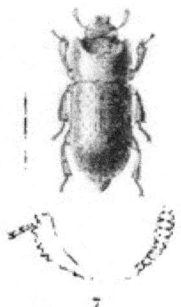

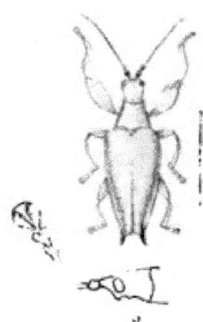

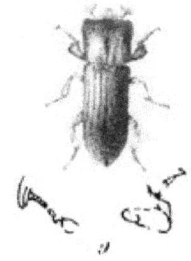

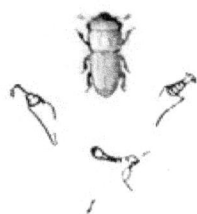

1

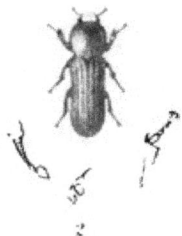

2

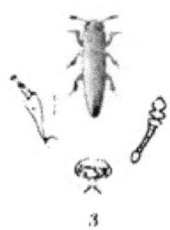

3

4

5

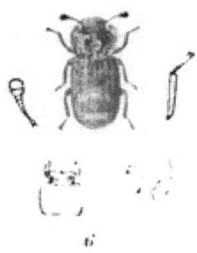

6

7

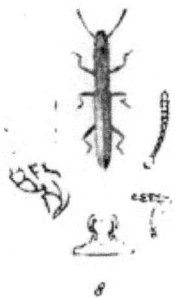

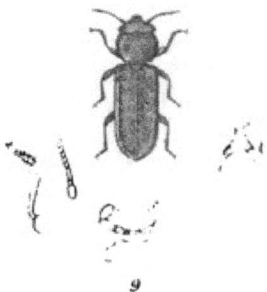

8

9

10

11

12

13

W.West. imp.

[*From the* JOURNAL OF ENTOMOLOGY, No. XIV., 1866.]

NOTICES

OF NEW OR

LITTLE-KNOWN GENERA AND SPECIES

OF

COLEOPTERA.

BY

FRANCIS P. PASCOE, F.L.S., ETC.

LATE PRESIDENT OF THE ENTOMOLOGICAL SOCIETY.

PLATES XVIII. & XIX.

PART V.

OCHROSANIS [Cucujidæ].

Caput rhomboideum, antice truncatum. *Labium* sat magnum, antice rotundatum. *Mentum* transversum, antice truncatum. *Palpi labiales* parvi. *Oculi* prominentes. *Palpi maxillares* validi, art. ult. securiformi. *Labrum* transverse quadratum. *Mandibulæ* bifidæ. *Antennæ* breves, claviformes*; *scapo* oblongo-ovato, art. 5 antennarum minus longiore; articulo secundo brevi, cæteris obconicis ad septimum sensim crassioribus, ultimo appendiculato duodecimum simulante. *Prothorax* elongato-quadratus, apice emarginatus, basi leviter lobatus. *Elytra* elongata, parallela, abdomine breviora. *Pedes* perbreves; *femora* compressa, ovata; *tibiæ* subfiliformes; *tarsi* heteromeri, paulo dilatati, art. penult. integro. *Coxæ* omnes subcontiguæ, posticæ processu interfemorali spiniformi separatæ. *Abdomen* segmentis æqualibus, ultimo paulo longiore excepto. *Corpus* elongatum, planatum, depressissimum.

The extraordinary flatness of this most singular insect in proportion to its size is, I should think, almost without a parallel among the whole class. Its nearest relationship at present appears to be with *Hemipeplus*, Latr., with which it agrees in a great number of

---

* This term seems in one quarter to have been misunderstood; I use it to denote "*club-formed*," *i. e.* gradually incrassated from the base to near the apex, where it contracts again,—in contra distinction to "*clavate*" or "*clubbed*," *i. e.* when there is a sudden enlargement or knobbing at the end.

2 K

particulars, but from which it essentially differs in the shorter basal
joint of the antennæ, and the great length of the elytra, which, notwithstanding, do not cover the abdomen.   *Hemipeplus* is known only
from a single specimen originally found in Scotland, and which has
been redescribed by M. Lacordaire; who, on the other hand, had not
seen the cognate genus *Inopeplus**, which, on reference to the first
volume of this work, Pl. XVI. fig. 9, will be seen to differ very decidedly in habit. I have dedicated the species to Mr. Dohrn of Stettin,
to whom I am indebted for my example.

Ochrosanis Dohrnii.   (Pl. XVIII. fig. 7.)

O. pallide ochracea; oculis nigris.
Hab. West Indies.

Pale ochre-yellow, very minutely punctured, and having an exceedingly delicate pubescence above; eyes black; prothorax rather concave
towards the base, with a strongly marked fovea near the posterior angle;
scutellum transverse, the sides at first parallel, triangular behind; elytra
broader than the prothorax, four times as long as broad, but leaving the
last and part of the penultimate segment of the abdomen uncovered:
abdomen beneath smoky brown, minutely piloso-granulated, sterna
ochraceous, smooth; antennæ nearly as long as the head and prothorax together.   Length 4½ lines.

Exarsus [Colydiidæ].

Caput retractum. *Oculi* subrotundati. *Antennæ* 11-articulatæ, subpilosæ,
clava triarticulata. *Palpi maxillares* sensim crassiores. *Mentum* quadratum. *Labium* valde transversum. *Maxillæ* lobo interiore apice
hamato. *Prothorax* medio elevatus, lateribus dilatatus, apice fortiter
sinuatus. *Elytra* convexa, rugosa, subquadrata, apice late rotundata.
Pedes modice elongati; *tibiæ* filiformes, inermes, ciliatæ; *tarsi* graciles,
articulis tribus basalibus æqualibus, subtus pilosis. *Corpus* amplum,
rugosum, marginibus ciliatum.

Allied to *Rechodes*, Er., but the prothorax and elytra ciliated at
their margins, not serrated, the tibiæ also ciliated, and the maxillary
palpi scarcely securiform. The genus contains one of the finest
species among the Colydiidæ, and is perhaps even more like *Asida*
and *Byrsax* among the Heteromera than *Rechodes*. I owe my specimens to the kindness of Robert Bakewell, Esq.

* = *Ino*, Lap. A name previously used by Leach for a genus of Moths, and
which has been recently revived. Mr. F. Smith, in his Catalogue of Cucujidæ of
the British Museum, long ago proposed to substitute " *Inopeplus.*" I have recently
seen an Australian example of this genus: it was sent as a *Staphylinus.*

Enarsus Bakewellii. (Pl. XIX. fig. 1.)

E. fuscus, squamosus, indumento terreno-griseo tectus.
Hab. New Zealand.

Dark brown, closely covered with a greyish or brownish-grey secretion, and with short, erect, more or less scattered scales; head deeply immersed in the prothorax, forming a nearly continuous line with the dilated margins of the latter; prothorax with a double gibbosity above the head, the dilated margins with two deep pits on each side; scutellum round; elytra slightly margined, deeply foveate, the suture raised in the middle, posteriorly abruptly declining to the apex, the declivity with three large callosities on each side; body beneath and legs with a dull rusty-brown tomentum. Length 4 lines.

ENNOMETES [Rhipiceridæ].

Caput antice brevissimum, labio minuto. *Oculi* magni, prominentes. *Palpi* acuti. *Antennæ* 11-articulatæ; *scapo* modice elongato, curvato, art. 2o obconico, reliquis paulo elongatis, flabellatis. *Prothorax* triangularis, apice truncatus, angulis posticis depressus, basi bisinuatus. *Elytra* prothorace haud latiora, angustata, subparallela. *Tibiæ* sublineares; *tarsi* filiformes, lamellis nullis, articulo ultimo cæteris simul sumptis breviore; *onychium* distinctum. *Coxæ* anticæ et intermediæ approximatæ, valde elongatæ. *Abdomen* segmentis quinque, basali brevissimo. *Corpus* angustum, fere parallelum.

The longer joints of the antennæ of this genus, contrasted with the very short joints in the allied form *Callirhipis*, although apparently not a very decided character, give to those organs such a very different appearance as to necessitate their separation; but in addition to this the tarsi are long, filiform, and with the last joint much shorter than all the preceding together. *Arrhaphus*, Kraatz, differs most essentially from *Callirhipis* and this genus in the three intermediate joints of the tarsi being bilobed, and very distinctly lamellated beneath. I find, in all the specimens of *Callirhipis* I have examined, only five abdominal segments—not six, as stated.

Ennometes Lacordairei. (Pl. XIX. fig. 2.)

E. ferrugineus; elytris dense seriatim punctatis, interstitiis paulo elevatis.
Hab. Queensland.

Ferruginous, slightly shining, very sparsely pubescent; antennæ cinnamon-brown; head and prothorax closely punctured, the latter with a slight horseshoe depression at the base; scutellum small, *circular*; elytra closely seriate-punctate, the intervals slightly raised; body beneath yellowish brown, rather glossy; legs clothed with stiffish hairs, mixed on the tibiæ with short spinous tubercles. Length 5 lines.

Psacus [Rhipiceridæ].

Caput verticale, retractum, antice triangulare. *Oculi* prominuli, rotundati. *Antennæ* 11-articulatæ; *scapo* subgloboso; art. 2 brevi; 3 trigono, ampliato; sequentibus flabellatis; tuberibus antenniferis nullis. *Palpi maxillares* robusti, art. ult. ovali. *Maxillæ* bilobæ, lobis ciliatis. *Prothorax* marginatus, transversus, basi bisinuatus. *Elytra* oblonga. *Pedes* breves; *femora* robusta; *tibiæ* lineares; *tarsi* filiformes, ciliatæ, lamellis nullis, articulo ultimo sine onychio. *Coxæ* anticæ transversæ, haud exsertæ, approximatæ, intermediæ retractæ. *Acetabula* antica magna. *Prosternum* angustum. *Abdomen* segmentis quinque, primo brevissimo.

A second specimen in my collection, which I have very little doubt is the female, differs from the above in its larger size, and the antennæ considerably less flabellate, the third joint slender and cylindrical, the fourth and fifth shorter and trigonate, the latter transversely, so as to make a beginning to the flabellate structure of the remainder. The male individual described above has been, unfortunately, so thickly gummed on the card, that I have had great difficulty in making out the underparts, and have been unable to obtain the lower lip and its palpi. It will be seen, from the description, that this genus fails in two characters hitherto considered essential to the family, viz. the absence of an onychium to the last tarsal joint, and the non-exserted anterior and intermediate coxæ. Nevertheless the antennæ are so entirely conformable that I think there can be little hesitation as to its being a real, although an aberrant, member of the group. The male has a strong resemblance to *Attagenus pellio*; the female I had put aside as a *Dermestes*. If either exists in collections, it will probably be found stowed away among the Dermestidæ.

Psacus attagenoides. (Pl. XVIII. fig. 4.)

P. oblongus, niger, villosus, indistincte fulvo marmoratus vel maculatus. *Hab.* South Australia (Gawler).

Oblong, black, sparsely covered with short erect hairs, and obscurely mottled or spotted with fulvous red; antennæ and legs ferruginous, except the black basal joint of the former; scutellum triangular; elytra obsoletely striated. Length 2 lines (♀ 3 lines).

Cnecosa [Telmatophilidæ].

Caput verticale, antice subtriangulare. *Oculi* prominuli, rotundati. *Antennæ* ante oculos insertæ, 11-articulatæ, art. 1 subgloboso, 2 breviore, 3 longiore, 4–8 subturbinatis, cæteris clavam magnam oblongam efficientibus, ultimo maximo. *Palpi maxillares* art. ultimo amplissimo, valde transverso. *Maxillæ* lobis duobus subæqualibus, ciliatis. *Labium* minutum. *Mentum* triangulare, apice late truncatum. *Palpi labiales* incrassati, art. duobus basalibus transversis, ultimo obconico. *Pro-thorax* transversus, lateraliter marginatus, apice paulo productus, basi ad elytra arcte applicatus, subbisinuatus. *Elytra* brevia, parallela, prothorace paulo latiora. *Pedes* breves; *femora* crassa; *tibiæ* modice elongatæ, subtrigonæ; *tarsi* subpentameri, æquales, articulis tribus basalibus crassis, transversis, quarto minuto, quinto cæteris simul sumptis longiore, unguiculis simplicibus. *Coxæ* anticæ ovatæ, haud approximatæ, intermediæ globosæ, distantes. *Prosternum* quadratum. *Mesosternum* declive, postice bilobatum. *Epimera* metathoracica parallela, postice truncata. *Abdomen* segmentis quinque subæqualibus. *Corpus* fere parallelum.

The subpentamerous tarsi, in the absence of any other striking character, appear to me to indicate the place of this genus to be among the Telmatophilidæ, the genera of which are not, however, very obviously connected, except by the above character.

Cnecosa fulvida. (Pl. XVIII. fig. 2.)

C. oblonga, clare fulva, leviter pubescens; oculis nigris.
Hab. New South Wales.

Oblong, clear fulvous-yellow, with a short, sparse, stiffish pubescence; head and prothorax finely punctured, the latter with an impressed line close to its base; scutellum transverse, slightly contracted at the base; elytra moderately seriate-punctate, with two rows of minute punctures between them; body beneath golden yellow, finely punctured. Length 2½ lines.

Antrisis [Scarabæidæ].

Caput transversum, verticale, clypeo inflecto. *Oculi* rotundati, sub angulis anticis prothoracis occulti. *Antennæ* 9-articulatæ; *scapo* elongato, curvato; art. 2 cylindrico, crasso; 3 obconico; 4, 5, 6 transversis; 7, 8, 9 lamellatis. *Mentum* antice rotundatum, in medio emarginatum. *Labium* membranaceum, ciliatum, minutum. *Palpi labiales* cylindrici, breves. *Maxillæ* lobo externo triangulari, interno membranaceo. *Palpi maxillares* elongati, art. ult. elongato securiformi. *Prothorax* transverso-quadratus, longitudinaliter carinatus. *Elytra* prothorace haud latiora, carinata, oblonga; pygidio obtecto. *Pedes* mediocres; *femora* antica et intermedia grossa, postica fusiformia; *tibiæ* tenuatæ, apice paulo dilatatæ, haud serratæ; *tarsi* lineares, postici longiores. *Coxæ* posticæ distantes. *Abdomen* segmentis 6, penultimo majore.

Apparently very closely allied in habit and characters to *Ryparus*, Westw., but differing from that and all the other genera of Apho-diinæ in the widely separated posterior coxæ. My example was kindly presented to me by W. Wilson Saunders, Esq.; another is in Mr. Wallace's Collection.

Antrisis Saundersii. (Pl. XVIII. fig. 5.)

A. griseo-fuliginea, punctata; antennis palpisque ferrugineis.

Hab. Sarawak.

Entirely greyish fuliginous; palpi and antennæ rusty; head with an impressed circular line in front, surrounded with eight tubercles; prothorax with eight strongly marked carinæ, the broad intervals irregularly punctured, the second carina on each side, counting from the two middle, with a deep linear oblique excavation anteriorly; scutellum punctiform; elytra with ten carinæ alternating with those on the prothorax, the intervals with two rows of punctures on each, the three intermediate carinæ on each elytron interrupted near the apex by a deep curved excavation, in which is placed a smooth, yellowish tubercle; body beneath and femora rather roughly punctured. Length 2¼ lines.

Intybia [Telephoridæ].

Caput antice triangulare. *Oculi* prominentes, ad angulos laterales positi. *Palpi maxillares* robusti, subcylindrici; *palpi labiales* minuti, subfusiformes. *Labium* trapezoidale. *Mentum* quadratum. *Antennæ* ad angulum inferum insertæ, 10-articulatæ; *scapo* elongato-clavato; art. secundo maximo; cæteris sat brevibus, cylindricis vel apicem versus subobconicis. *Prothorax* capite angustior, apice paulo rotundatus, postice constrictus, basi truncatus. *Elytra* obovata, convexa, basi prothorace latiora. *Pedes* graciles; *tarsi* filiformes, 5-articulati, art. ult. trigono.

In habit this genus resembles the females of *Charopus*; but the antennæ and characters generally are those of *Collops*. The eyes are almost semipedunculate, as in the cognate form *Cephalogonius*.

Intybia guttata. (Pl. XVIII. fig. 6.)

I. nigra, genis flavis; elytris dilutioribus, singulis guttis tribus albis ornatis.

Hab. Batchian.

Black; head finely granulated, cheeks yellow; antennæ black, the undersides of the four or five basal joints yellowish; prothorax finely granulated, a deep, transverse, irregular depression near the base; scutellum transversely triangular; elytra paler or smoky black, finely pubescent, each with three distinct white spots, one near the base, and two towards the apex, the innermost approaching the suture; body beneath black, sides of the abdomen yellow. Length 1½ line.

As the greater part of the following belong to the Tenebrionidæ, the subfamilies (corresponding invariably to the "tribes" of M. Lacordaire) are given after their genera.

Adesmia* eburnea.

A. aterrima; elytris late ovatis, albis, tricostatis, costis remote dentatis. *Hab.* N'Gami.

Jet-black, shining, the elytra ivory-white; clypeus slightly emarginate; head finely punctured; prothorax impunctate, very transverse, the anterior angles not produced; elytra broadly ovate, very little longer than broad, dead ivory-white, obsoletely impressed, each with three very distinct but slightly elevated costæ, the two inner crowned with sharp, slender, distantly set teeth, the outer with a double row of more closely set and shorter teeth; body beneath and legs black; the abdomen and sterna finely corrugated. Length 4½ lines.

A remarkable species, very distinct from the other white-winged members of the genus (*Langei, candidipennis*, &c.) in the form of the elytra and their toothed costæ. It was taken by Mr. Anderson in South Africa, *somewhere* north of Natal, and towards Lake N'Gami.

DYSARCHUS [Asidinæ].

Caput transversum, *retractum*; *clypeus* fronte confusus, labrum et mandibulas obtegens. *Oculi* transversi, angusti. *Palpi maxillares* fortiter securiformes; *labiales* minuti. *Mentum* transverso-quadratum, angulis anticis rotundatis. *Antennæ* breves, 11-articulatæ; art. 3 longiore; 4–6 brevioribus, subquadratis; 7 breviter obconico; 8–10 transversis et compressis; 11 minore quam præcedens, rotundato. *Prothorax* transversus, ad latera rotundatus, apice semicirculariter emarginatus, basi truncatus, angulis posticis paulo productis. *Elytra* ovato-rotundata, prothorace latiora; epipleuræ basi latæ, postice sensim angustatæ. *Pedes* validi; *tibiæ* anticæ extus compressæ, infra emarginatæ, bidentatæ; posticæ et intermediæ trigonatæ, calcaratæ; *tarsi* infra biseriatim ciliati, intermedii et postici art. ultimo breviore quam primus. *Sterna* et *abdomen* ut in *Asida*.

The clypeus being confounded with the front, nearly hiding the lip and mandible, is a character at variance with the rest of the subfamily. The fore tibiæ are those of *Anomalipus* (placed by Solier in this group); the tarsi, closely ciliated on each side beneath, appear in consequence canaliculate. The granules with which the upper parts are covered rise abruptly out of a greyish exudation, and are very irregular in form and size.

Dysarchus Odewahnii.

D. obscure niger, granulis nitidis instructus. *Hab.* South Australia (Gawler).

Dull greyish black, covered above with numerous glossy granules;

---

* Fischer de Waldheim, Entomogr. de la Russie, i. 153; Lacordaire, Gen. v. p. 23.

head nearly flat anteriorly; antennæ ferruginous, scarcely half the length of the prothorax; the latter with a strongly marked margin at the sides, the granules giving it a serrated appearance at the edge; scutellum triangular, deeply sunk beneath the base of the prothorax; elytra with four prominent granular lines, more or less interrupted, on each, the lines towards the apex gradually disappearing, the intervals with smaller granules; epipleuræ roughly granulose; body beneath black, the abdomen glossy; legs closely punctured, clothed with a thin ferruginous pubescence; teeth on the fore tibiæ strongly produced, especially the apical. Length 5½ lines.

Emeax [Scaurinæ].

Caput subelongatum, collo incrassato. *Oculi* transversi, liberi. *Palpi* subcylindrici. *Mandibulæ* bifidæ. *Mentum* transversum, antice et lateribus rotundatum. *Antennæ* sublineares, art. secundo brevi, tertio breviore quam quartus, cæteris brevioribus et obconicis, ultimo minore quam præcedens. *Prothorax* subtransversus, apice et basi truncatus, lateribus rotundatis et fortiter carinatis, angulis posticis acutis. *Elytra* prothorace vix latiora, oblonga, lateribus leviter rotundatis. *Pedes* mediocres; *tibiæ* subtrigonatæ, bicalcaratæ; *tarsi* postici art. primo subelongato. *Prosternum* productum. *Mesosternum* elongatum, declive. *Processus* interfemoralis quadratus, apice late angulatus.

In the form of the head, the mentum entirely covering the maxillæ, leaving a little only of the lower lip visible, the small terminal joint of the antennæ, and the large intermediate acetabula, this genus approaches some of the Scaurinæ, and particularly in habit *Psamme-tichus*, Latr. I am not quite sure, however, that a better place may not be found for it eventually.

Emeax sculpturatus.   (Pl. XIX. fig. 7.)

E. niger, opacus; capite prothoraceque rugoso-punctatis, illo in medio cristato; elytris grosse tuberculato-lineatis.
Hab. New South Wales.

Black, opake; head coarsely and closely, the neck finely punctured, between the eyes a prominent tuberculiform crest; prothorax closely and roughly punctured, on the disk two large foveæ, the lateral carina crenate; scutellum depressed, triangular; elytra narrowed at the base, the shoulders rounded, each with eight lines of large oblong tubercles, in the interval lines of smaller tubercles; body beneath and legs dull blackish brown, closely punctured. Length 7 lines.

Ossiporis [Molurinæ].

Caput exsertum, verticale, antice quadratum et excavatum. *Oculi* parvi, rotundati, producti, vel quasi pedicellati. *Palpi max.* art. ult. obconico. *Labium* rotundatum. *Antennæ* attenuatæ, squamosæ, art. tertio elon-

gato, 4° 5°que subæqualibus, 6–9 sensim brevioribus, 10° transverso, 11°
globoso-ovato. *Prothorax* fere transversus, convexus, lateraliter angulatus, antice paulo productus, basi sinuatus, disco æquatus. *Elytra*
ovata, basi prothorace haud latiora, supra subplanata; *epipleuræ* sub-
angustatæ. *Pedes* graciles; *tibiæ* teretes; *tarsi* post. art. basali quam
ultimus longiore. *Prosternum* postice curvatum. *Mesosternum* declive.

The head of this curious insect has a marked resemblance to that
of the hippopotamus. The eye is surmounted by a projecting orbit,
which causes it to protrude in such a way as to give it the appearance
of being almost pedicellate. Below the eyes the face is concave,
and is particularly deeply excavated between the antennary orbits.
The genus is related to *Phligra*.

Ossiporis terrena.

O. supra indumento terreno tecta, infra pedibusque squamulis albidis
densissime vestita; antennis squamosis, articulis duobus terminalibus
nigris.
Hab. Natal.

Covered above with an earthy crust, composed of flattish scales and
short projecting points; legs and body beneath entirely covered with a
uniform layer of flat whitish scales; lower part of the head and lip, and
antennæ, clothed with loose whitish scales, the two last joints of the
antennæ black. Length 5½ lines.

Onosterrhus [Pedininæ].

Affinis *Pedino*, sed *oculi* non divisi. *Mentum* subcordiforme. *Prothorax*
margine laterali limbo replicato. *Tibiæ* sensim latiores, haud trigonatæ.
Tarsi postici art. basali cæteris simul sumptis fere æquali. *Corpus* valde
convexum.

The fold bordering the upper part of the edge of the prothorax is
also characteristic of *Trigonopus*; but the fore tibiæ of that genus
are even more triangular than those of *Pedinus*, and it also differs in
most of the above characters. In its general appearance the species
described below resembles *Heliopathes Lusitanicus*, but is larger and
much more convex.

Onosterrhus lævis.

O. niger, subnitidus; capite subtiliter punctato; prothorace elytrisque
impunctatis.
Hab. Western Australia.

Black, slightly nitid; head finely punctured, concave anteriorly;
prothorax impunctate, much broader than the head, well-rounded at the
sides, but a little incurved at the base, the posterior angle slightly produced, the lateral margins with a conspicuous uniform fold bordering

its edge, and creating a strongly marked groove on its inner side;
scutellum short and very transverse; elytra impunctate, shortly ovate,
broader than the prothorax, to which they are closely applied, the
shoulders rounded: epipleuræ broad at the base, gradually narrowing to
the apex; body beneath and legs smooth and somewhat glossy; antennæ as long as the prothorax, the 8th, 9th and 10th joints transverse,
11th rounded.   Length 6 lines.

Idisia [Opatrinæ?].

Caput porrectum, subelongatum, ad oculos retractum.   *Labium* valde
transversum.   *Oculi* rotundati, prothoraci approximati.   *Palpi maxillares* cylindrici.   *Antennæ* robustæ, ciliatæ, clavatæ, 11-articulatæ,
scapo crasso, art. 3o longiore, cæteris brevibus, clava triarticulata, art.
ult. minore.   *Prothorax* transversus, lateribus rotundatus, ciliato-mar-
ginatus, apice truncatus, basi bisinuatus.   *Scutellum* invisum.   *Elytra*
ovata, costata; epipleuræ postice angustiores.   *Pedes* mediocres; *femora*
incrassata, trochanteribus intermediis nullis; *tibiæ* anticæ trigonatæ,
cæteris linearibus calcaratis; *tarsi* lineares, antici breves, art. basali brevissimo, intermedii et posteriores elongati.   *Coxæ* anticæ globosæ, distantes.   *Episterna metathoracica* linearia, epimeris propriis obsoletis.
Prosternum elevatum, latum.   *Mesosternum* declive.   *Metasternum* breve.
Processus interfemoralis mediocris, antice rotundatus.   *Corpus* gracile,
squamulosum.

Having only a single specimen of this insect, for which, and an
extensive collection of Coleoptera made on the coast of Chinese
Tartary, I am indebted to Arthur Adams, Esq., R.N., I have not
attempted to extract its oral organs; but, judging solely from the
characters that remain, I do not see that it can well be referred to
any of the numerous groups described by M. Lacordaire.   The habit
in some respects suggests *Stenosinæ*; but the clavate antennæ, ciliated
tarsi, spurred tibiæ, and retracted head are at variance with that subfamily.   In its scaly clothing it is similar in character to *Leichenum
pulchellum*, but more delicate; and this, in conjunction with its tarsi
and trigonate anterior tibiæ, induces me to refer it, although doubtfully, to Opatrinæ.

Idisia ornata.   (Pl. XVIII. fig. 8.)

I. nigra, squamulis albis tecta; elytris basi ochraceis, in medio fascia
grisea irregulari ornata.
Hab. Mantchuria.

Black, entirely covered by small white scales: lip glabrous, brown;
eyes with subspinous facets, placed at a little distance behind the antennary orbits; antennæ not larger than the prothorax, reddish brown,
but with scattered white scale-like hairs; prothorax with a central
impressed line: elytra with five elevated lines on each, the first,

second, and fifth or external, only, extending to the base, where they form strongly marked projections, and have an ochreous colour, the middle of the elytra with a darkish irregular band ; body beneath black, with scattered setaceous white hairs ; legs pale brown, with white scale-like hairs. Length 2¼ lines.

Nyctobates* Orcus.

N. niger, nitidus ; prothorace lævigato, lateribus vix rotundatis, postice angulatim constrictis; elytris punctato-striatis ; prosterno lato, tricarinato; tarsis validis.

Hab. Western Australia.

Black, shining ; head and prothorax impunctate, the latter nearly as broad as the elytra, trapezoidal, the sides anteriorly slightly rounded, then nearly parallel to near the base, where it contracts at an angle ; scutellum triangular, below the level of the basal ridge of the elytra ; elytra ovate, scarcely broader than the prothorax at the base, punctate-striate, the punctures large, the striæ shallow ; body beneath glossy black; prosternum broad, marked with three strong rounded ridges, the lateral not united behind ; legs glossy brownish black, tarsi stout, closely covered with bright-yellowish-ferruginous hairs. Length 12 lines.

Differs from *N. crenatus*, Boisd., in the form of the prothorax, the punctate-striate elytra, stouter tarsi, and the prosternum strongly tricarinated throughout.

Nyctobates feronioides.

N. niger, nitidus ; prothorace subtilissime punctulato, lateribus rotundatis, postice incurvato-constrictis; elytris punctato-striatis ; prosterno angustato, leviter marginato ; tarsis validis, brunneo-castaneis.

Hab. New South Wales.

Black, shining ; head and prothorax very minutely punctured, the latter narrower than the elytra, fully rounded at the sides, contracted and a little incurved posteriorly ; scutellum triangular, continuous with the basal ridge of the elytra; elytra oblong, broader at the base than the prothorax, the shoulders a little recurved, punctate-striate, the punctures rather small, but the striæ deep ; body beneath glossy black ; prosternum narrowed, pointed behind, slightly margined ; legs brownish black, glossy ; tarsi stout, brownish chestnut. Length 7 lines.

Besides the smaller size, this species differs from *N. crenatus*, *inter alia*, in the deeper striæ, more strongly, and not crenately, punctured elytra, and stouter tarsi.

* Guérin-Méneville, Mag. de Zool. 1834, p. 33 ; Lacordaire, Gen. v. p. 371. This genus and the three following should have been placed after the *Bolitophagina*.

Toxicum* punctipenne.

I. nigrum, opacum; cornibus capitis elongatis; prothoracis apice lobato; elytris fortiter lineato-punctatis; antennarum clava triarticulata.

Hab. Australia.

Black, opake; head deeply excavated between the four horns, and coarsely and remotely punctured, the posterior horns with a fulvous tuft on the apex; eyes undivided; labrum black; antennæ with the last three joints only forming the club; prothorax moderately transverse, finely punctured, anterior angles scarcely produced, broadly lobed at the apex; scutellum triangular; elytra broader than the prothorax, coarsely punctured in lines; body beneath and legs black, shining. Length 2¼ lines.

Near *T. quadricorne*, Fab., but with the club of the antennæ with three joints only, a narrower prothorax, and coarsely punctured elytra.

Toxicum brevicorne.

T. nigrum, opacum; cornibus capitis brevibus; prothoracis apice haud lobato; antennarum clava triarticulata.

Hab. Victoria.

Black, opake; head slightly excavated in front, moderately but more closely punctured; posterior horns short, triangular, with the small tuft of fulvous hairs confined to the anterior part of the apex; lip testaceous-brown; club of the antennæ four-jointed; prothorax slightly transverse, not lobed at the apex, finely punctured; scutellum triangular; elytra lineate-punctate, the punctures of moderate size; body beneath and legs chestnut-brown, shining. Length 4 lines.

A very distinct species, but agreeing with the above in having the eyes undivided: this character and the number of joints (three or four) composing the club seem in this genus to be of secondary value only. I have seen no female either of this or the preceding.

Uloma† depressa.

U. depressa, pallide rufo-ferruginea; elytris striato-punctatis, punctis distinctis sed parvis; tibiis anticis extus quadridentatis.

Hab. Queensland.

Rather broad, depressed, pale reddish or orange-ferruginous, but sometimes much darker; head finely punctured, a slight curved impression in front; prothorax smooth, glossy, minutely punctured; scutellum small, scutiform; elytra striato-punctate, the junctures distinct but rather small; body beneath and legs brownish orange, shining; fore tibiæ with four stout serratures externally towards the apex, the intermediate with five or six small serratures, the posterior smooth on both sides. Length 6 lines.

---

* Latreille, Gen. Crust. et Ins. ii. p. 167; Lacordaire, Gen. v. p. 341.

† Laporte de Castelnau, Hist. Nat. des Ins. ii. p. 219; Lacordaire, Gen. v. p. 332.

This species is broader and much more depressed than *Uloma culinaris*, Linn., and the fore tibiæ are rather serrated than toothed; the punctures on the elytra are also very much deeper and more distinct.

Dechius [Tenebrioninæ].

Tenebrioni affinis, sed differt præcipue *labro* obtecto. *Maxillæ* lobo interno mutico. *Antennæ* articulis 8–10 transversis. *Prothorax* fortiter marginatus, marginibus antice productis.

The characters of *Tenebrio*, like those of perhaps the majority of the genera of Coleoptera, were only exposed in a really definite way in the 'Genera' of M. Lacordaire. Compared with his description, the above formula differentiates the two genera. The habit, moreover, is somewhat different, being more convex and cylindrical, and sufficiently suggestive of the *Aphodius*-form to justify the specific name proposed.

Dechius aphodioides.

D. fusco-ferrugineus, subnitidus; capite prothoraceque fusco-nigris; elytris fortiter striato-punctatis.

Hab. Queensland.

Brownish ferruginous, subnitid; head and prothorax brownish black, the edges of the clypeus reddish ferruginous, finely punctured, the prothorax grooved along the base, the raised margin bordering the groove depressed in the middle, where it is joined by the nearly obsolete median line; scutellum pentagonal; elytra a little broader than the prothorax at the base, strongly striato-punctate, the margins well marked, especially at the shoulders; body beneath dark brown, shining, the breast rufescent; legs reddish ferruginous; antennæ extending to about the middle of the prothorax, reddish brown. Length 4½ lines.

Scymena [Trachyscelinæ].

Characteres ut in *Phaleria*, sed *clypeus* profunde quadrato-excisus. *Antennæ* capite breviores. *Processus* interfemoralis apice acutus.

I have examined only the oral organs *in situ*; but they appear to be pretty nearly of the same character as those of *Phaleria*. The type described below resembles *P. Gayi*, Lap. Of my two examples one is reddish testaceous, the other black.

Scymena variabilis.

S. rufo-testacea vel nigra, nitida; elytris fortiter punctato-striatis, interstitiis striarum subtiliter punctulatis.

Hab. Australia.

Shortly ovate, reddish testaceous or black, shining; head finely punctured; clypeus separated from the front by a well-marked semilunar line; antennæ shorter than the breadth of the head, imperfoliate; prothorax finely punctured, the apex rather strongly emarginate; scutellum broadly triangular; elytra punctate-striate, the striæ sharply defined, the intervals minutely punctured; body beneath and legs dull testaceous or black; tibiæ and tarsi roughly ciliated. Length $2\frac{1}{2}$-3 lines.

Ecripsis [Trachyscelinæ].

Characteres ut in *Ammobio*, sed *palpi maxillares* securiformes; *antennæ* longiores. *Tarsi* articulo ultimo elongato.

The securiform palpi are an exception to the rest of the Trachyscelinæ, except *Sphærgeris*; the unusual length of the last tarsal joint will, however, distinguish this genus from all the others of the subfamily.

Ecripsis pubescens.

E. rufo-testacea, disperse griseo-pubescens.
Hab. Tasmania.

Shortly ovate, reddish testaceous, with a pubescence composed of short scattered hairs; head covered with small crowded granules; clypeus separated from the front by a well-marked semicircular line; antennæ as long as the breadth of the head; prothorax minutely and closely granulose; scutellum very broadly triangular; elytra nearly impunctate, the sides with the pubescence more setose and elongate; body beneath and legs reddish testaceous, sparingly pubescent; tibiæ granulose, the anterior with the outer apical portion triangular, with a comparatively slight emargination above. Length $1\frac{3}{4}$ line.

Isarida [Trachyscelinæ].

Oculi detecti. *Prothorax* basi lobatus. *Tibiæ* intermediæ et posticæ lineares, ciliatæ; *tarsis* propriis elongatis. *Prosternum* declive, haud lanciforme, mesosterno distinctum. Cæteris ut in *Ammobio*, sed *corpus* minus convexum.

From *Anemia* this genus, like *Ammobius*, will be distinguished, *inter alia*, by its retractile anterior tarsi: the principal characters separating it from the latter lie in the four posterior tibiæ and tarsi, and the prosternum.

Isarida testacea.

I. fulvo-testacea, glabra, subnitida; elytris subtiliter granulato-punctatis.
Hab. India (Dacca).

Fulvous-testaceous, glabrous, subnitid; head finely punctured; clypeus nearly confounded with the front, angularly emarginate; length of the antennæ scarcely half the width of the head; prothorax finely punctured, very transverse, lobed at the base; scutellum broadly triangular; elytra finely granulato-punctate, bordered, as well as the prothorax, with long bristly hairs at regular intervals; body beneath and legs fulvous, rather regularly punctured; anterior tibiæ strongly trigonate, the emargination above the external apical angle very deep and rounded, the other tibiæ coarsely ciliated. Length 1½ line.

Hyocis [Trachyscelinæ].

Caput insertum, antice modice elongatum, clypeus haud distinctus. *Oculi* prominuli, rotundati. *Labrum* transversum, apice integrum. *Palpi maxillares* art. ult. cultriformi. *Antennæ* prothorace breviores, subperfoliatæ; art. primo longiore; tertio haud elongato; 2–8 breviter obconicis, majoribus, transversis; ultimo orbiculari. *Prothorax* elytris haud contiguus, transversus, apice late emarginatus, lateribus rotundatus, postice angustior, pone angulum posticum constrictus, basi truncatus. *Elytra* ovata, convexa. *Pedes* mediocres; *tibiæ* anticæ triangulares, omnes extus spinosulæ; *tarsi* lineares, postici art. basali modice elongato. *Prosternum* haud productum. *Mesosternum* declive. *Processus* interfemoralis latus, antice rotundatus.

The type of this genus is a little insect having somewhat the general appearance of a *Cryptophagus*, and which appears to me to come between *Ammobius* and *Phaleria*. I have not, however, ventured to examine the oral organs of my solitary specimen, for which I am indebted to Mr. Bakewell; but apparently they are not very different from those of *Phaleria*.

Hyocis Bakewellii.

H. fusco-ferruginea; elytris fortiter striato-punctatis.
Hab. Victoria.

Dark ferruginous; head and prothorax closely punctured, the punctures rather large and shallow, the raised intervals having a reticulate appearance; antennæ reddish testaceous, enlarging in thickness from the fifth to the tenth joint; prothorax with an impressed median line, its anterior angles rounded, its posterior acute; scutellum small, triangular; elytra scarcely broader than the prothorax, coarsely striato-punctate or almost clathrate; sterna covered with close shallow punctures; on the abdomen they are scattered and nearly obsolete, each with a small silvery hair; legs reddish testaceous. Length 1¼ line.

Ozolais [Bolitophaginæ].

Characteres capitis, oculorum &c. ut in *Ilyxero*. *Palpi labiales* art. ult. ovato, obtuso. *Labium* latum, haud emarginatum. *Antennæ* 10-articulatæ, clavatæ; *scapo* elongato, modice incrassato; art. 2–8 subcylindricis, robustis; duobus ultimis clavam validam formantibus (9 semicirculari, 10 rotundato). *Prothorax* gibbosus, angulis anticis productis, marginibus crenatis. *Elytra* subcylindrica, prothorace paulo latiora. *Pedes* ut in *Ilyxero*. *Prosternum* productum, in cavitate Λ formante mesosterni receptum. *Corpus* oblongum, tuberculatum.

The antennæ of this genus are also clavate; but then they are ten-jointed, as in *Bolitotherus*. The specimen described below is one of Mr. Bates's discoveries on the Amazons.

Ozolais scruposa. (Pl. XVIII. fig. 1.)

O. fusca, pube grisea induta.
Hab. Ega.

Dark brown, with a loose greyish pile; head with two rows of tubercles in front; prothorax very convex or gibbous above, with nine or ten crenatures on each side, the penultimate by far the largest and directed backwards, the base rather strongly emarginate for the reception of part of the scutellum; elytra seriate-punctate, with several large tubercles mixed with others smaller and more spinous; body beneath dark brown, closely punctured; legs and antennæ reddish brown, with scattered grey hairs. Length 2 lines.

ILYXERUS [Bolitophaginæ].

Caput ante oculos dilatatum; clypeus distinctus. *Oculi* magni, semidivisi. *Palpi maxillares* robusti, apice oblique truncati. *Palpi labiales* art. ult. ovali, acuto. *Labium* parum emarginatum. *Mentum* transversum, antice sinuatum, postice constrictum. *Antennæ* 11-articulatæ, clavatæ; *scapo* elongato, crasso; art. 2 brevi; 3 obconico; 4–8 breviter trigonatis; tribus ultimis clavam validam compressam formantibus (9 valde transverso, 10 et 11 arcte applicatis, rotundatis). *Prothorax* postice constrictus, angulis anticis productis; marginibus crenatis, haud foliaceis. *Elytra* elongata, parallela, prothorace haud latiora. *Pedes* mediocres; *tibiæ* subcylindricæ, inermes; *tarsi* lineares, art. ult. cæteris simul sumptis vix breviore. *Prosternum* productum. *Mesosternum* subverticale. *Metasternum* modice elongatum. *Corpus* parallelum, angustatum, tuberculatum.

The narrow parallel form of this genus would at once distinguish it from any other in its subfamily; and no other has yet been published with clavate antennæ. I owe my specimen to Mr. Bakewell's liberality.

Ilyxerus asper. (Pl. XVIII. fig. 3.)

I. griseo-fuscus, supra tuberculatus et punctatus.
Hab. New South Wales.

Brown, with a sparse greyish-ochreous pubescence; head with a large tubercle over each eye, and three smaller ones between them; prothorax with about eight crenatures on each side, the disk roughly tuberculate. Scutellum semicircular; elytra with six rows of tubercles on each, between each row two lines of well-marked punctures; body beneath dark brown, the legs and antennæ reddish brown, all clothed with scattered greyish setulose hairs. Length 2½ lines.

Byrsax* Macleayi.

B. oblongus, fuscus; capite maris cornibus elevatis, apicem versus incurvatis et decussatis; prothorace tuberculato, disco 4-tuberculato;
elytris subdisperse punctatis, tuberculis magnis subseriatim positis.
Hab. Australia.

Oblong, dark brown, opake; head of the male armed with two long
stout vertical horns, incurved and crossing each other at the tips, the
tips themselves emarginate or reduced in thickness; head of the female
with a simple tubercle between the eyes; prothorax finely punctured,
very tuberculate, four principal tubercles on the disk towards the base
arranged in pairs (::), in the female two others at the extreme apex,
much produced, slightly recurved and transversely compressed; elytra
somewhat coarsely punctured (the punctures rather depressed), tuberculate, four principal tubercles oblong and very large, on each side of the
suture; between these and the margin on each side three slightly irregular rows of smaller and rounder tubercles; body beneath and legs
reddish brown; mesosternum with a very compressed vertical process;
club of the antennæ 7-jointed. Length 5 lines.

The genus *Byrsax* was proposed by me in the first number of this
Journal (April, 1860), and differs from M. Motschoulsky's *Boli-
toxenus* ('Études, &c.,' 1858, p. 63), in that the elytra have a produced
margin, which is always coarsely serrate, and the prosternum is keeled
anteriorly. *Byrsax* was there, in consequence of its tarsi appearing
to me to be tetramerous, referred to the Colydiidæ; at the same time
I pointed out its resemblance to *Diaperis horrida*, Ol. (a true *Byrsax*),
but stated that, " guided by its tetramerous tarsi," its real affinity
would be with *Endophlœus, Pristoderus*, and some other genera. I
am now satisfied that it is truly heteromerous, the basal joint, indeed,
being completely hidden in the cotyloid cavity of the tibia. I am not
so satisfied, however, that the resemblance between it and the above
Colydiide genera is *only* one of analogy. *Bolitophagus gibbifer*, Wes-
mael, is possibly identical with *Byrsax cœnosus*. There are, however, several other undescribed species.

Byrsax egenus.

B. oblongus, indumento terrulento fulvescente tectus; prothorace gibboso,
disco 8-calloso, callis tuberculatis; elytris subseriatim callosis.
Hab. Australia.

Oblong, covered with a fulvescent tomenticious substance; head with
four tubercles between the eyes, and two on the clypeus; prothorax
very gibbous, the disk with four large callosities anteriorly, each apparently made up of three or four conical tubercles, and four smaller

ones behind, all arranged in pairs, the sides rugosely tuberculate; elytra
with comparatively few tubercles, placed in three irregular rows, the
principal tubercle at the shoulder; body beneath brown; legs and
antennæ reddish chestnut; mesosternum produced anteriorly. Length
2 lines.

My specimen appears to be a female, but in colour, size, &c. it differs considerably from the above. I received these two species from
Mr. MacLeay, but without any special locality. They are I think,
most interesting additions to the fauna of that island continent.

Ceropria peregrina.*

C. nigra, nitida; prothorace subtilissime punctato, basi utrinque fortiter
impresso; elytris striato-punctatis, punctis confertis; tarsis ferrugineis.

Hab. Queensland.

Black, shining; head finely punctured, clypeus extending to the eyes,
truncate in front: antennæ with the joints from 4-10 equal, equilaterally triangular, 11 orbicular; prothorax minutely punctured, a large
fovea on each side at the base: elytra punctate-striate, the punctures
squarish and very close together, the spaces between the striæ finely
punctured; body beneath pitchy; legs glossy black, the tarsi filiform,
ferruginous. Length 4½ lines.

EMYPSARA [Trachyscelinæ].

Caput subretractum; clypeus rotundatus; *labium* transversum. *Oculi*
transversi, prothorace subobtecti. *Antennæ* mediocres, articulis apice
ciliatis, *scapo* oblongo, art. 2 brevi, 3 obconico et longiore, 4, 5, 6 breviter obconicis, 7, 8, 9, 10 transversis, ultimo rotundato. *Palpi maxillares*
robusti, art. ult. ovato, præcedente haud latiore. *Maxillæ* lobo interno
angusto, hamato. *Palpi labiales* parvi, remoti, labii basi externe inserti.
Labium transversum, antice late emarginatum, postice contractum.
Mentum transverse quadratum. *Prothorax* transversus, antice angustior, basi truncatus, lateraliter marginatus. *Elytra* prothorace latiora,
brevia, epipleuræ angustatæ. *Tibiæ* trigonatæ, calearatæ. *Tarsi* antici
et intermedii art. 2º et 3º dilatatis, postici filiformes. *Prosternum*
postice productum, procossu in fovea excisa mesosterni recepto. *Corpus*
globoso-ovatum, marginibus ciliatum.

Nearly all the characters of this genus are, in the main, similar
to those of *Phaleria*, except the ciliated margin of the body and the
dilated anterior and intermediate tarsi—in the former approaching
Trachyscelinæ and *Anemia*, belonging to the same subfamily, and in
the latter the Pedininæ. The elytra are finely striated: nine striæ
may be counted on each; they are therefore rather widely apart. I

* Laporte de Castelnau et Brullé, Ann. des Sc. Nat. xxiii. p. 306; Lacordaire,
Gen. v. p. 307.

have not been able to ascertain if these insects are winged. In *Phaleria cadaverina* there are only the rudiments of wings. The two species were kindly presented to me by Arthur Adams, Esq.

Empsura Adamsii. (Pl. XIX. fig. 3.)

E. nigra ; elytris, sutura nigra excepta, testaceis, fasciis flexuosis duabus rufis ornatis.

Hab. Mantchuria (Vladimir Bay).

Black, the upper surface minutely and closely granulated ; antennæ shorter than the length of the prothorax, pale at the base ; scutellum broadly triangular ; elytra testaceous, except the black suture and two broad waved reddish bands ; body beneath and legs dark chestnut-brown. Length 3 lines.

Fig. 2*a* represents the fore tarsus, and fig. 2*b* the antennæ.

Empsura flexuosa.

E. testacea ; capite supra nigro ; elytris lineis flexuosis nigro notatis.

Hab. Mantchuria (Oo-oo Bay).

Testaceous, the upper surface minutely and closely granulated ; head black above ; antennæ darker at the tips, longer than the length of the prothorax ; scutellum triangular ; elytra with two series of irregular patches, forming two imperfect bands, more brownish testaceous than the rest, and bordered more or less with dark-brown or black wavy lines ; body beneath and legs yellowish testaceous, the spurs of the tibiæ and claws black. Length 2½ lines.

Pterohelæus pruinosus.*

P. breviter ovatus, fuscus, pulvere albido tectus ; elytris striato-punctatis, singulatim costis tribus vix elevatis instructis.

Hab. North Australia.

Allied to *P. piceus*, Kirby, but broader, and the sides more paralllel, covered with a fine uniform whitish exudation, and, under the lens, a scattered greyish squamosity ; elytra striate-punctate, with only three very slightly raised lines on each ; body beneath reddish chestnut ; antennæ and legs ferruginous. Length 9 lines.

Pterohelæus agonus.

P. ovatus, fuscus, subnitidus ; prothorace apice late emarginato ; elytris tenuiter lineato-punctatis, lineis subremotis.

Hab. South Australia.

Ovate, blackish brown, slightly nitid ; head very finely punctured ; antennæ and palpi ferruginous ; prothorax nearly impunctate, very short, broadly emarginate at the apex, the posterior angles slightly

---

* De Brême, Essai Monog. et Iconog. de la Tribu des Cossyphides, p. 27 : Lacordaire, Gen. v. p. 346.

produced; scutellum subtriangular; elytra lineate-punctate, the punctures small, the lines rather widely apart; body beneath and legs black, shining; tarsi ferruginous.   Length 5–6 lines.

At once distinguished from *P. striato-punctatus*, Bois., and *P. Kollari*, de Brême, by the broad semicircular emargination of the apex of the prothorax.   From the former, which it more nearly resembles, it may also be known by the nearly impunctate prothorax, and the elytra more decidedly lineate-punctate, not irregularly punctured on the disk.   Dr. Boisduval's name is apt to mislead, as there are no striæ.

Pteroheloeus servus.

P. oblongus, glaber, obscure fuscus; prothorace apice anguste sed profunde emarginato, in medio linea impressa; elytris prothorace paulo angustioribus, striato-punctatis, striis approximatis.
Hab. Victoria.

Narrower than *P. silphoides* *, De Br., with the prothorax a little wider than the elytra, its apex more deeply and squarely emarginate—not semicircular—and the narrowly impressed line in the middle more strongly marked; elytra striato-punctate, the striæ approximate; body beneath and legs glossy chestnut-brown; sides of the abdominal segments wrinkled.   Length 7 lines.

Pteroheloeus memnonius.

P. oblongus, glaber, niger, subnitidus; capite angusto; oculis magnis, subapproximatis; prothoracis marginibus subtiliter corrugatis; elytris lineato-punctatis.
Hab. South Australia (Adelaide).

Oblong, glabrous, black, slightly nitid; head finely punctured, narrowed; the eyes large and subapproximate, the distance between them in front being rather more than the length of their shortest diameter; prothorax finely punctured, its margins minutely waved; elytra closely lineate-punctate, the punctures well marked, the margins very narrow; body beneath and legs black, shining; tarsi and lip with ferruginous hairs.   Length 11 lines.

A large species resembling *P. silphoides*; but the narrow head, large eyes, and fine waved lines on the margins of the prothorax will differentiate it from all its congeners.

Pteroheloeus bullatus.

P. angusto-oblongus, rufo-brunneus vel fuscus; prothorace subtiliter punctato; elytris submulticostatis, costis granulatis.
Hab. South Australia (Queensland).

---

* M. Lacordaire puts this species under *Saragus*; but it is winged, with the usual correlation of a long metasternum.

Narrowly oblong, reddish brown or dark brown, slightly shining; head finely punctured; prothorax with very minute punctures, the emargination at the apex very shallow; elytra rather finely lineate-punctate, the alternate lines slightly elevated (about nine on each elytron) and garnished with small glossy pustular or bubble-like granules placed at irregular intervals on those lines; body beneath dark chestnut-brown, or paler; legs also varying from reddish to brown, and shining. Length 8 lines.

Apparently allied to *Cilibe granulosus*, De Br., from New Zealand?, but essentially different in the very minute, almost obsolete punctation of the prothorax. It varies in colour, but the margins of the prothorax are paler than the disk.

Helæus consularis.*

H. obovatus, niger, nitidus, marginibus latis valde reflexis; prothorace in medio dentato-carinato, dente posteriore magno spiniformi; scutello carinato; elytris impunctatis, marginibus exceptis, utroque costa crenata et linea tuberculata externa instructo.
Hab. Western Australia.

Obovate, glabrous, black, shining; prothorax with a toothed carina in the middle, the posterior tooth large, in the form of a compressed triangular spine; scutellum transversely triangular, keeled in the middle; elytra impunctate, the broad strongly reflexed margins faintly punctured, their edges terminated by an erect, narrow border, the raised suture having on each side at a short distance a sharp costa crenated on both sides, and externally near the angle formed by the reflected margin a line of small tubercles which do not, however, extend to the base; body beneath glabrous, black, shining; legs rugose. Length 11 lines.

About the average shape and size of *H. colossus*, but with elytral costæ as in *H. Peronii*. The following is very nearly allied.

Helæus moniliferus.

H. ovatus, glaber, brunneus, nitidus, marginibus latis, explanatis; scutello semicirculari, haud carinato; elytris fere obsolete punctatis, cæteris ut in *præcedente.*
Hab. South Australia.

Broader and paler than the last, the margins scarcely reflexed; the scutellum rounded posteriorly, or semicircular, without a keel; elytra with the lateral costæ less crenated, and the exterior line of tubercles extending to the base, and the punctation, though minute, very evident under the lens. Length 11 lines.

---

* Latreille, Règne An. ed. 1, iii. p. 301; Lacordaire, Gen. v. p. 347.

Helæus castor.

H. late ovatus, fusco-brunneus, vix nitidus, setuliferus, marginibus latis, explanatis; prothorace angulis posticis productis, incurvis; elytris sutura elevata, utroque costa valida usque ad partem tertiam percurrente.

Hab. South Australia.

Broadly ovate, brownish, scarcely shining, margins of the prothorax and elytra broad, and only slightly reflexed, the edges with an erect, narrow border; antennæ nearly linear; prothorax with short, scattered bristly hairs, the perforated portion with an elevated margin, disk with a sharp longitudinal line, posterior angles falcate, overlapping the elytra; scutellum broadly triangular; elytra irregularly punctured, and clothed with numerous scattered minute bristles, costa on each side the suture terminating at about a third from the apex; body beneath dark brown; legs rugose. Length 9 lines.

A broad, stout species, the sides of the elytra within the margins more vertical and elevated than in any other.

Helæus falcatus.

H. ovatus, niger; prothorace marginibus anticis in processum acute falcatum terminatis; elytris lividis, sparse setuliferis, in utroque costa crenata usque ad tertiam partem percurrente.

Hab. Kangaroo Island.

Differs from *H. Peronii,* Bois., in the following particulars:—antennæ narrower, much less dilated at the apex; prothorax obsoletely granulous, its anterior processes gradually narrower to the point, or, in other words, completely falcate, not of equal width until near the point, and not hollowed out above; elytra very glossy, as if varnished, their surface very slightly punctured and with scattered *curved* bristly hairs, and the lateral costa more decidedly crenate. Length 5–6 lines.

SYMPETES [Helæinæ].

Characteres ut in *Helæo,* sed *caput* liberum, anguli antici prothoracis haud producti. *Labrum* obtectum.

The form of the prothorax will not allow the species described below being placed in *Helæus*; and the declivous mesosternum without any notch for the reception of the prosternal process, independently of other characters, separate it from *Saragus*. It is quite an *Helæus* in habit. I received my specimen from Mr. MacLeay; and it is the only one I have seen. A second species has been described by Mr. White, under the name of *Encephalus tricostellus* (App. to Gray's Voyage, p. 461).

Sympetes Macleayi.

S. late ovatus, valde depressus, latissime marginatus, fuscus ; elytris confertim punctulatis, singulis subtrilineatis.

Hab. Australia.

Broadly and almost elliptically ovate, and very much depressed, dark **brown, the margins paler,** somewhat shining ; clypeus broad, emarginate at the apex, and hiding the lip; prothorax finely punctured, the disk at the base scarcely more than a third of the width ; scutellum broadly triangular ; elytra with rather small punctures, sharply raised along the suture, each elytron with three indistinct lines, the margins nearly flat ; body beneath and legs dull reddish brown, the margins of the prothorax and elytra finely punctured. Length 12 lines.

Saragus* magister.

S. elliptico-ovatus, impunctatus, niger, nitidus ; prothorace elytrisque eat fortiter marginatis.

Hab. Queensland.

Elliptic-ovate, black and shining, very smooth and impunctate ; clypeus transverse, gradually rounded from the antennary orbits ; prothorax rather narrowly emarginate at the apex, the disk moderately convex, the margin about one-sixth of the breadth of the disk at its widest part ; elytra convex, slightly raised into a line posteriorly at the suture, the margins narrower than those of the prothorax ; body beneath glossy black, the abdomen finely corrugated ; femora highly polished ; tibiæ and tarsi with fulvous hairs, the latter and the antennæ ferruginous. Length 9 lines.

As regards sculpture and outline this species will come into the same category of the genus as *S. brunnipes,* De Br. ; but it is much larger, and at once differs in the entire absence of punctation.

Saragus asidoides.

S. elliptico-ovatus, niger, opacus ; prothorace lateribus dilatato, subtiliter et confertissime punctulato; elytris lineato-punctatis.

Hab. South Australia (Adelaide).

Elliptic-ovate, black, opake ; clypeus slightly produced and truncate in front ; head finely punctured ; prothorax minutely and very closely punctured, rather narrowly emarginate at the apex, the disk flattish and passing gradually into the margin on each side, shining ; elytra flattish at the base, more convex posteriorly, with small punctures in slightly irregular lines, the margin very distinct at the shoulders, but gradually narrowing to the apex, where it nearly disappears ; body beneath and legs black, slightly nitid ; the latter and antennæ with a thin ferruginous pubescence. Length 7 lines.

Something like *Asida depressa,* Sol., but more convex. The second

* Erichson, Wieg. Arch. 1842, i. p. 171 ; Lacordaire, Genera, v. p. 318.

and third joints of the anterior tarsi are somewhat dilated in my specimen.

Saragus Duboulaii.

S. subrotundus, niger, opacus, late marginatus; elytris confertim impresso-punctatis.

Hab. Western Australia (Champion Bay).

Nearly round, brownish black, opake, covered with a very short brownish pubescence; clypeus very broad, truncate or very slightly emarginate, with a deep groove in the middle behind at its junction with the head; prothorax nearly impunctate, narrowly and deeply emarginate at the apex, the disk scarcely convex, less than half the breadth at the base, and separated from the margins by a strongly marked curved impression; elytra with numerous small punctures, the disk slightly concave, the margin very distinct, gradually narrowing posteriorly; body beneath like the upper part; legs and antennæ pale reddish brown, with a short greyish pile. Length 5 lines.

One of the flattest and most nearly circular of the whole subfamily. I owe my example to Mr. Duboulay.

Saragus exulans.

S. oblongo-ovatus, fusco-brunneus, subnitidus; prothorace vix marginato; elytris confertim lineato-punctatis, marginibus angustatis.

Hab. Lord Howe's Island.

Oblong-ovate, convex, dark reddish brown, subnitid; clypeus sloping at the sides, the apex emarginate; head finely but rugosely punctured; prothorax closely and finely punctured, the punctures here and there confluent, the apex broadly emarginate, the lateral margins nearly confounded with the disk; elytra closely lineate-punctate, the margins very narrow, and almost disappearing posteriorly; body beneath and femora glossy chestnut-brown; tibiæ, tarsi, and antennæ pale ferruginous, finely pubescent. Length 5½ lines.

Resembles *S. brunnipes*, but without any dilatation of the margin of the prothorax, &c.

Saragus infelix.

S. breviter ovatus, fuscus, opacus; prothorace confertim punctato; elytris tricostatis, interstitiis vage punctatis; tibiis scabris.

Hab. Tasmania.

Shortly ovate, blackish brown, opake; clypeus very transverse, narrower anteriorly and emarginate, separated from the head by a deep semicircular line; head finely but rugosely punctured; prothorax closely covered with small oblong punctures, the margins broad and subgranulous, the apex rather broadly emarginate; elytra tricostate, the costæ moderately elevated, dying out towards the apex, the intervals irregularly punctured, the margins narrow but very distinct;

body beneath and femora chestnut-brown; tibiæ scabrous, the outer edge of the anterior tuberculate; tarsi and antennæ pale ferruginous. Length 6 lines.

Allied to *S. lævicollis*, Fab., but less convex, the disk of the prothorax not granulate, the intercostal spaces simply punctured, and the tibiæ covered with small asperities.

Saragus Odewahnii.

S. breviter ovatus, fuscus, opacus; prothorace confertissime punctulato; elytris tricostatis, costis duabus externis interruptis, interna postice abrupte abbreviata, interstitiis subtiliter granulosis; tibiis subscabris.
Hab. South Australia (Gawler).

Shortly ovate, blackish brown, opake; clypeus slightly produced, broadly emarginate at the apex, separated from the head by an indistinct semicircular line; head rugosely punctured; prothorax very closely and minutely punctured, the intervals having a granulous appearance, the apex rather broadly emarginate, the margins broad and pale brownish; elytra considerably broader at the middle and posteriorly, tricostate, the inner costa sharply defined, but suddenly ceasing before the apex, the two outer costæ broken up into short lines or points, the intervals with a slight tomentose pubescence, out of which rise a number of minute granules; body beneath dark brown, shining; legs paler; tibiæ slightly scabrous. Length 5 lines.

Resembles the preceding, but smaller, the elytra more decidedly narrowed at the base; the sculpture of the prothorax and elytra are also very distinctive. It was one of the many novelties sent to me by Mr. Odewahn.

OSPIDUS [Helæinæ].

Caput ad oculos retractum; clypeo distincto, integro. *Oculi* transversi, supra distantes. *Palpi maxillares* securiformes, *labiales* breviter claviformes. *Maxillæ* lobo externo rotundato, interno angusto, inermi. *Labium* ampliatum, apice emarginatum. *Mentum* transverso-sexangulare. *Antennæ* 11-articulatæ, art. tertio longiore, reliquis gradatim brevioribus et crassioribus, quatuor ultimis clavam compressam formantibus, ultimo rotundato. *Prothorax* transversus, lateribus rotundatus et angusto-marginatus, apice profunde emarginatus, basi subbisinuatus. *Elytra* late ovata, convexa, leviter marginata; epipleuræ excavatæ. Alatæ. *Pedes* breviusculi; *tibiæ* anticæ apice extus productæ, inermes; tibiæ intermediæ et posticæ bicalcaratæ; *tarsi* subæquales, art. primo duobus sequentibus longiore. *Prosternum* carinatum, postice productum. *Mesosternum* Λ-forme pro processu prosterni. *Metasternum* modice elongatum. *Processus* interfemoralis triangularis. *Corpus* convexum, breviter ovatum.

This genus is allied to *Cilibe* as limited by M. Lacordaire. It has,

however, a somewhat elongate metasternum, a character which that
learned entomologist may probably consider inconsistent with such
affinity.

Ospidus chrysomeloides.

O. fusco-æneus; elytris cupreis, subtiliter punctatis; antennis fuscis.
Hab. Queensland.

Dark brownish bronze; head closely punctured; lip broad, scarcely
emarginate; prothorax very minutely and closely punctured, the sides
with small vermicular elevations; scutellum broadly triangular; elytra
copper-brown, with numerous small punctures and several indistinct
longitudinal raised lines; body beneath and legs reddish copper, shining;
antennæ dark brown. Length 6 lines.

Cossyphus* Odewahnii.

C. ovatus, modice elongatus, testaceus, limbo subtilissime reticulato;
prothorace haud carinato; elytris striato-punctatis, insertionis linea
secundum elytra biseriatim punctata.
Hab. South Australia (Gawler).

Ovate, slightly elongate, testaceous, rim subdiaphanous, the reticulations exceedingly delicate, and scarcely visible under the lens, except
in certain lights; prothorax without any carina, almost obsoletely
punctured, darker than the rim; scutellum triangular; elytra striato-
punctate, the punctures rather coarse and irregular, the junction of the
rim with the disk marked with two rows of large punctures; body beneath, legs, &c. testaceous. Length 2 lines.

The punctures in two rows at the junction of the disk with the
rim of the elytra, although large, are not very well defined; but they
are distinctive, and differentiate this species from any of those described by the Marquis de Brême, none of which, however, are from
Australia. Since his essay, M. Peyron and Dr. Gerstaecker have
each published one, from Syria and Mozambique respectively. I
have two more undescribed from India.

Eutelus† ovatus.

E. niger, indumento terrulento griseo tectus; prothorace medio bicalloso;
elytris ovatis, tuberculis numerosis irregulariter dispersis.
Hab. Natal.

Black, covered with a thick greyish or reddish-grey secretion mixed
with short hairs; clypeus short, transverse, front of the head with a
few glossy black tubercles; prothorax transversely subglobose, in the

* Olivier, Entom. iii. No. 44 bis, p. 3.

† Solier, Mem. Accad. Torino, ser. 2, vi. p. 268; Lacordaire, Gen. v. p. 355.

middle two longitudinal crests crowned by numerous small tubercles, at the sides also two smaller round prominences placed directly midway between the apex and base; elytra ovate, rather broader than the prothorax, covered with several (about a dozen or more on each elytron) oblong callosities running parallel to but at a distance from the suture, and with a few minute granules between them; body beneath and legs black, shining, and rugose, with a short pubescence partially filling in the hollows; tarsi glabrous, glossy black. Length 6–7 lines.

This is probably the "second" species alluded to by M. Lacordaire (Gen. v. p. 356). It differs from Soher's *E. nodosus* in its ovate, not globose, elytra, each with a dozen or more oblong longitudinal elevations or tubercles, not with six only, "joined three and three, and forming two transverse callosities." M. Lacordaire's description of *Eutelus* applies perfectly to this species and to *E. nodosus*; but from some oversight, he has represented as the latter the species I have described below under the name of *Cyrtotyche Satanas*.

CYRTOTYCHE [Eutelinæ].

Characteres ut in *Eutelo*; sed *caput* retractum. *Elytra* epipleuris antice obsoletis. *Tibiæ* elongatæ, tenuatæ, valde curvatæ, anticæ intus productæ. *Corpus* glabrum.

A remarkable form, closely allied to *Eutelus*, but which, owing to the absence of the peculiar vestiture of that genus, the longer legs, and slender, curved tibiæ, has quite another habit. It is figured on M. Lacordaire's plate under the name of *Eutelus nodosus*, Sol.

Cyrtotyche Satanas.

C. purpureo-fusca, nitida; antennis, palpis tarsisque nigris.
Hab. Natal.

Dark purplish brown, shining; head minutely punctured; prothorax rather broader than the elytra, very convex and arched above, and much contracted at the base, four glossy oblong tubercles in a transverse line in the centre, with two on each side behind, the uppermost very large and prominent, the lowermost bifid, the intervals rather dull, irregular and impunctate; scutellum very small, triangular; elytra impunctate, short and globose, scarcely as broad as the prothorax at the base, irregularly covered with large conical tubercles, the intervals pitted here and there, especially near the suture; body beneath black, shining; legs dark purple, shining; tarsi black. Length 7 lines.

BYZACNUS [Eutelinæ].

Characteres ut in *Eutelo*; sed *prothorax* subquadratus, supra æqualis, modice convexus, lateribus paulo rotundatus, apice late emarginatus, carina

laterali distinctus.   *Elytra* elevata, a medio fortiter declivia, epipleuris
linearibus.   *Tibiæ* lineares, subcurvatæ.   *Corpus* glabrum.

These characters give this genus a very different appearance from
Eutelus and the preceding one.   The form of the prothorax causes a
marked distinction between the pronotum and its flanks, which is
wanting in the above genera; the epipleuræ of the elytra are also
of equal breadth throughout, and very apparent.

Byzaenus picticollis.   (Pl. XIX. fig. 6.)

B. cupreo-fuscus, nitidus; prothorace lateribus rubris; antennis, tibiis
tarsisque rufo-ferrugineis.
Hab. Natal.

  Copper-brown, shining; head finely punctured; antennæ reddish ferruginous; lip and palpi paler; prothorax rugosely punctured, the sides
dark red; scutellum small, transversely convex; elytra coarsely and
sparingly punctured, narrower at the base than the prothorax, gradually
broader and higher near the middle, where they are a little rounded and
furnished with twelve tubercles, then almost vertically declivous, and
the sides rapidly narrowing to the dehiscent apex, where they terminate each in an obtuse point; body beneath and femora dark brown,
shining, finely punctured; tibiæ and tarsi reddish ferruginous.   Length
5 lines.

Oremasis [Cyphaleinæ].

A *Prophane* differt *prothorace* angulis anticis haud productis; elytris
gibbosis, basi prothorace haud latioribus et ad eum arcte applicatis,
apicibus acuminatis.

The type of this genus is *Adelium cupreum*, G. R. Gray (Griffith's
Animal King. Ins. ii. p. 22, pl. 80. f. 2).   It is another of those
species allied to *Prophanes*, which M. Lacordaire considers not to
accord with that genus.   It is in fact much nearer *Cyphaleus* in
form than the depressed *Prophanes*.   Another genus to be constituted is one for *P. metallescens*, Westw.; and probably my *Chartopteryx binodosus*, with its two great humps like *Thecacerus*, will
require another.   All the species of this handsome family, except
*Lepispilus sulcicollis** and *Hemicyclus grandis*, are exceedingly rare.
My collection contains eleven species, the British Museum seven, the
greater part of each of each of them limited to a single individual.
There are doubtless many more to come.

Lygestira [Cyphaleinæ].

Prophani affinis, sed differt *clypeo* emarginato; angulis anticis prothoracis
haud productis; et apicibus elytrorum muticis.

* Found in Australia as well as in Tasmania.

This is one of the forms, mentioned by M. Lacordaire, referred to *Prophanes*, Westw., but having characters essentially different. They would probably, however, be considered slight, if they were not accompanied by a marked difference of habit, and the species did not belong to a family in which the instability of characters does not always admit of very trenchant differentiation. *Prophanes simplex*, Westw., belongs to this genus: my specimen scarcely answers in some particulars to Mr. Westwood's description : but it agrees in having a *deeply* emarginate clypeus, a character by which it differs from the species described below ; the eyes are also much more widely apart, and the prothorax at the apex is considerably broader than the head, the anterior angles being consequently well marked though not produced. It is one of the many good things sent me by Mr. Odewahn.

Lygestira funerea.

L. olivaceo-nigra, nitida; oculis parvis, fronte remotis; clypeo modice emarginato ; prothorace apice late emarginato.

Hab. South Australia (Gawler).

Oblong-ovate, depressed, olive-black, or very dark olive, shining, and having a very thin brownish tomentum ; head and prothorax minutely punctured ; the latter broadly emarginate at the apex, the anterior angle on each side extending considerably from the head ; eyes small, widely apart; elytra finely punctured, longitudinally marked with numerous almost obsolete costæ, the apex rounded ; body beneath black, and legs black, shining, tibiæ finely ciliated. Length 11 lines.

CHOLIPUS [Cnodalinæ].

Caput exsertum, antice transversum ; *clypeus* distinctus, fere ad basin antennarum truncatus. *Labrum* transversum. *Mandibulæ* apice integræ. *Palpi maxillares* securiformes ; *labiales* art. ult. magno, cyathiformi. *Mentum* trapeziforme, in linea mediana incrassatum. *Antennæ* art. 3io quam 1mus longiore, 6–10 sensim latioribus transversis et compressis, ultimo præcedente majore et rotundato. *Prothorax* subtransversus, apice fere truncatus, ad latera rotundatus, tenuiter marginatus, postice constrictus, basi leviter bisinuatus. *Elytra* prothorace latiora, oblonga, pone medium latiora. *Tibiæ* subcurvatæ, muticæ ; *tarsi* art. ultimo majore. *Prosternum* dilatatum, declive. *Mesosternum* antice excavatum.

The type of this genus has long been known under Dejean's name here adopted. With it I have associated a species from Queensland which appears to me to be congeneric. The genus is distinguished *inter alia* from its allies by its declivous prosternum without the usual mucro.

Cholipus brevicornis.

C. niger, nitidus; elytris purpurascentibus, leviter striato-punctatis, punctis subtilissimis interstitiis sitis.

Hab. Java.

Glossy black; head and prothorax finely punctured, the latter with a median impressed line; scutellum triangular; elytra dark purplish, finely striato-punctate, the interstices with extremely minute punctures; body beneath and legs glossy black; antennæ with the last six joints but one dilated on one side. Length 6 lines.

Cholipus punctipennis.

C. niger, subnitidus; elytris seriatim punctato-impressis, punctis mediocribus, interstitiis impunctatis.

Hab. Queensland.

Black, slightly shining; head and prothorax minutely punctured; the latter with the median impressed line confined to the posterior part, and ending in a shallow fovea near the base; scutellum triangular; elytra seriate-punctate, the punctures impressed, moderately large, the intervals impunctate; body beneath, legs, and antennæ glossy black. Length 6 lines.

Hemicyclus* punctulatus.

H. breviter elliptico-ovatus, purpureo- vel chalybeato metallicus; elytris subtiliter punctatis.

Hab. South Australia.

Shortly ovate-elliptical, steel-blue, varying to purple: head sparsely punctured, a broad longitudinal impression between the eyes; antennæ chalybeate blue, the basal joint copper; prothorax nearly impunctate, except at the rim; scutellum triangular; elytra finely punctured, the punctures numerous but not crowded; body beneath and legs dark greenish black, shining. Length 6½ lines.

Very decisively distinguished from *H. grandis* and *metallicus* of Westwood by its oblong although still somewhat hemispherical form and punctured elytra.

Platyphanes† cyaneus.

H. ovatus, cyaneus, nitidus: capite prothoraceque nigris: antennis pedibusque subferrugineis.

Hab. North Australia.

Ovate, very convex, deep indigo-blue, shining; head and prothorax black, minutely punctured; clypeus not distinct from the head, slightly emarginate; antennæ scarcely longer than the breadth of the head, ferruginous, the 7th–10th joints transverse, the last as long as broad; prothorax much narrower than the elytra, broadly lobed at the base; elytra gradually broader to about the middle, then rounding to the apex,

---

lineate-punctate, the punctures very small and in about 14 lines, but the fifth and eighth forming double lines of smaller punctures, all disappearing near the apex ; body beneath shining ; femora nearly black ; tibiæ and tarsi subferruginous. Length 8 lines.

Probably a **proper genus.** I have referred it here on account of its antennæ, **rather than** to *Cyphaleus*, but the epipleuræ of the elytra are not horizontal as in *Platyphanes*. Is not *Platyphanes vittatus*, **Westw., an** *Hemicera* ?

EUCYRTUS [Cnodalinæ].

Caput exsertum, ante oculos brevissimum; clypeus distinctus, transversus, ad basin antennarum truncatus, apice emarginatus. *Labrum* transversum. *Mandibulæ* apice integræ. *Palpi maxillares* securiformes ; *labiales* art. ultimo elongato-ovato. *Mentum* haud carinatum, transversum, antice truncatum. *Palpi labiales* art. ult. oblongo-ovato. *Antennæ* art. 3io quam 1mus longiore, 5o vel 6–10 sensim latioribus et brevioribus, compressis, ultimo majore, rotundato. *Prothorax* transversus, apice subsinuatus, ad latera rotundatus, fortiter marginatus, postice constrictus, basi bisinuatus. *Elytra* late ovata, prothorace latiora. *Pedes* validi ; *femora* intermedia et posteriora aliquando infra hirsuta ; *tibiæ* rectæ, muticæ, intus ad apicem villosæ ; *tarsi* subdilatati, articulo ultimo cæteris simul sumptis breviore. *Prosternum* latum, postice depressum, mucrone brevi verticali terminatum. *Mesosternum* antice excavatum. *Processus* interfemoralis triangularis.

These characters apply to *Eucyrtus pretiosus* of Dejean's catalogue, briefly described by M. Lacordaire in his 'Genera' (v. p. 417, *note*), but they do not apply to Hope's *Scotæus*, into which that learned entomologist has merged the genus of Dejean. *Scotæus* has a retracted head, the clypeus produced beyond the antennary lobes, the antennæ subserrate, prothorax finely margined at the sides, **the four posterior femora not hairy beneath,** the tibiæ **glabrous ; the tarsi are much longer** and more **attenuated, with the claw-joint** very long ; **above all, the** prosternum is horizontally prolonged posteriorly, and is **received into** a long narrow groove of the mesosternum. Borneo **seems to be rich** in the **species of** *Eucyrtus*.

GAUROMAIA [Cnodalinæ].

Characteres ut in *Eucyrta*, sed *clypeus* productus, orbitis antennarum distinctus, emarginatus. *Maxillæ* lobo interiore dentato. *Mentum* fere semicirculare, basi in processum triangularem productum. *Palpi labiales* art. ult. breviter obconico. *Femora* sublinearia. *Tibiæ* antice teminatæ, curvatæ.

I am unwilling to introduce characters derived almost entirely from the oral organs as distinctive of nearly allied genera, principally because they are difficult to ascertain, and are probably assumed in the majority of cases, except in the type of the genus: but in this instance the toothed inner maxillary lobe is an exception to the characters of this subfamily as laid down by M. Lacordaire. The mentum is also very peculiar, it is very much produced in front in a nearly semicircular form, with short sides each of which gives off a sort of triangular process at right angles to the main axis, in this way forming a kind of enclosure round the base of the lower lip. The clypeus *let into* the front and distinct from the antennary orbits, is also very different from the same organ in *Eucyrtus*, where it runs into and seems a part of the orbits.

Gauromaia dives.

G. purpureo-metallica, nitida; corpore infra, antennis pedibusque nigris.
Hab. Malacca (Mount Ophir).

Metallic purple, smooth and shining; head and prothorax finely punctured, the latter transverse, rounded and finely margined at the sides and base, the latter and apex of nearly equal breadth; scutellum triangular; elytra ovate, broader than the prothorax, striatopunctate, punctures small and approximate, spaces between the striæ broad and minutely punctulate; body beneath and legs black; femora not thickened, the anterior slightly curved, a faint shade of coral-red in the middle; tibiæ nearly equal in length, more or less curved; tarsi subequal; antennæ black, with the last four joints pubescent. Length 7 lines.

PHAEDIS [Cnodalinæ].

Characteres ut in *Eucyrto*; sed *clypeus* productus; *mentum* minus transversum, in medio carinatum. *Antennæ* breviores, articulis magis transversis. *Prothorax* apice paulo productus. *Femora* incrassata, in medio subtus dentata.

The last character also distinguishes this genus from the foregoing, from which, moreover, it differs in the larger comparative size of the claw-joint, and the shortness of the anterior tibiæ.

Phaedis elysius.

P. cæruleus, nitidus; elytris aureo-æneis, lineato-punctatis.
Hab. Sarawak.

Sky-blue, very smooth and glossy; head rather remotely punctured, the lateral line, separating the clypeus from the front, extending above the base of the former; antennæ reddish ferruginous, pubescent, scarcely longer than the breadth of the head; palpi pale ferruginous; prothorax with numerous small, rather remote punctures, finely bordered on all sides; scutellum small, triangular; elytra much broader than the prothorax, bright brassy yellow, or brassy with a golden tinge, lineate-punctate, the intervals between the lines with very minute scattered points, the intervals between the punctures brassy brown; body beneath and legs glossy, with a greenish tinge. Length 5 lines.

ELIXOTA [Cnodalinæ].

Caput breve, usque ad medium oculorum insertum; *clypeus* angustatus, apice truncatus. *Labrum* productum. *Oculi* magni. *Mentum* trapezoidale, medio carinatum, apice emarginatum. *Labium* subcordatum. *Palpi labiales* approximati, in medio labii inserti, art. ult. amplo. *Antennæ* apicem versus paulo incrassatæ, articulis obconicis, ultimo ovato. *Prothorax* transversus, lateribus rotundatus, apice emarginatus, basi sublobatus. *Elytra* prothoraci arcte applicata, oblonga, convexa; *epipleuræ* integræ. *Pedes* mediocres; *tibiæ* graciles: *tarsi* lineares, articulo ultimo mediocri. *Prosternum* postice productum, in incisura mesosterni receptum. *Processus* interfemoralis triangularis. *Corpus* oblongo-ovatum.

Allied to *Damatris*, Lap., but differentiated by its small claw-joints, and the epipleuræ of the elytra entire and gradually narrowing to the apex. The last joint of the maxillary palpus is nearly three times the size of the outer maxillary lobe. The type resembles an oblong *Chrysomela* (*C. geminata* for instance), only much larger.

Elixota cuprea.

E. cuprea, nitida; elytris lineato-punctatis, interstitiis punctulatis.
Hab. North China.

Dark copper-brown, shining; head and prothorax finely punctured, the junction of the clypeus with the front marked only by a deep transverse line, which does not extend to the antennary orbits, the six terminal joints of the antennæ broader than the others; scutellum triangular, its sides slightly rounded; elytra oblong, rather broader than the prothorax, lineate-punctate, the punctures rather small, oblong, and occasionally approximate, the intervals between the lines remotely and minutely punctate; body beneath dark copper-brown, shining; posterior tarsi with the basal joint nearly as long as the rest together, the claw-joint not longer than the second and third together. Length 4½ lines.

NAUTES [Cnodalinæ].

Caput retractum; *clypeus* fronte confusus. *Labrum* transversum. *Mandibulæ* apice integræ. *Palpi maxillares* securiformes: *labiales* art. ult. magno, cyathiformi. *Mentum* quadratum. *Labium* magnum, basi angustiore. *Antennæ* graciles, art. tertio longiore, 4-7 æqualibus tenuatis, 8-10 elongato-obconicis, ultimo oblongo oblique truncato. *Prothorax* transversus, apice leviter emarginatus, basi sublobatus. *Elytra* oblonga, prothorace paulo latiora: *epipleuræ* postice abrupte angustiores.

2 M

Pedes mediocres; *tibiæ* lineares; *tarsi* antici art. tribus basalibus dilatatis (♂?) tertio subbilobo, penultimo parvo : intermedii et postici art. basali paulo elongato, penultimo minore, ultimo mediocri.   *Prosternum* postice in cavitate mesosterni receptum.   *Corpus* oblongo-ovatum, convexum.

I have adopted this genus, which does not appear to have been published, from one of M. Deyrolle's lists.   It differs from the former, *inter alia*, in the third joint of the anterior tarsus being subbilobed, with the small penultimate joint inserted between the lobes.

Nautes fervidus.

N. cupreus, nitidus ; elytris lineato-sulcatis, sulcis fere obsolete punctatis. *Hab.* Mexico.

Bright reddish copper, shining ; head finely but not closely punctured; lip testaceous, hairy ; prothorax minutely and rather remotely punctured, the margin on each side thickened ; scutellum triangular ; elytra with nine narrow but deeply sulcated lines on each, the outermost not extending to the base, the lines marked at regular intervals as if punctured ; body beneath and legs dark greenish copper; tarsi and antennæ ferruginous.   Length 4 lines.

Arcothymus [Helopinæ].

Caput retractum; *clypeus* distinctus, antice rotundatus.   *Oculi* transversi, distantes.   *Palpi maxillares* securiformes ; *labiales* cylindrici.   *Mentum* transversum, antice bisinuatum, postice valde constrictum.   *Labium* apice emarginatum.   *Antennæ* filiformes ; articulo tertio elongato, cæteris longitudine æqualibus, subobconicis, ultimo oblongo-ovato. *Prothorax* transversus, antice rotundatus, pone medium incurvatus angulis posticis acutis, apice late emarginatus, basi truncatus.   *Elytra* prothorace basi latiora, dorso subplanata, lateribus subito declivia et rotundata, epipleuræ angustatæ.   *Pedes* inæquales, graciles ; *tibiæ* calcaratæ ; *tarsi* postici articulo basali cæteris simul sumptis fere longiore. *Prosternum* productum.   *Mesosternum* longiusculum, latum, subverticale.   *Processus* interfemoralis latus.

The type of this genus is rather above the ordinary size, and in outline resembles the shorter species of *Blaps.*   Like so many other genera in this family, it does not appear, to have any very obvious affinities.

Arcothymus cœnosus.

A. niger, pube scabra fuscescente tectus, infra glaber, nitidus ; elytris obsolete striatis.
Hab. Australia.

Black, covered with a rough brownish pubescence composed of short stiff hairs, mixed with a slight powdery exudation; head much narrower than the prothorax, retracted to the eyes; antennæ extending to nearly half the length of the body; prothorax very slightly convex, the inflected sides nearly three times the breadth of the epipleuræ, which are very distinct; body beneath black, shining; legs with a thin brownish pubescence. Length 8 lines.

MIMOPEUS [Helopinæ].

Caput retractum; *clypeus* fronte confusus, apice emarginatus. *Oculi* transversi, distantes. *Palpi maxillares* securiformes; *labiales* triangulares. *Mentum* antice latum, basin versus angustatum. *Antennæ* breviusculæ, art. tertio longiore, 4–7 brevioribus obconicis, 8–10 latioribus et compressis plus minusve obconicis, 11 rotundato. *Prothorax* transversus, ad latera rotundatus, antice angustior, basi bisinuatus, apice fortiter emarginatus. *Elytra* ovata, prothorace latiora, humeri angulis prothoracis obtecti. *Pedes* mediocres; *tibiæ* apicem versus crassiores, calcaratæ; *tarsi* antici breves, postici elongati tenues. *Prosternum* productum. *Mesosternum* breve, triangulare. *Processus* interfemoralis quadratus, antice rotundatus.

In the character of its antennæ this genus is nearer to *Misolampus* and its allies than to *Helops*, with which it agrees in habit. In this respect, however, it still more nearly resembles some species of *Amara*.

Mimopeus amaroides.

M. glaber; capite prothoraceque nigris, nitidis; elytris purpureo-brunneis, punctato-impressis, subtilissime granulosis; antennis pedibusque ferrugineis.

Hab. Australia.

Smooth; head and prothorax black, shining, finely punctured, lateral margins of the latter slightly produced, bordered with a slightly raised edge; scutellum transversely triangular; elytra dark purplish brown, opake, with numerous irregular punctiform impressions, the intervals dotted with very minute granules, epipleuræ ferruginous, broad at the base, gradually narrowing towards the apex; body beneath dark purplish brown, shining; legs, antennæ, and palpi ferruginous. Length 6 lines.

GNESIS [Helopinæ].

Caput subverticale, antice dilatatum et integrum; *labrum* emarginatum, basi angustius. *Oculi* transversi, sinuati. *Palpi maxillares* art. ult. securiformi. *Maxillæ* lobo interno hamato. *Palpi labiales* art. ult. ovato. *Labium* integrum, basi angustius. *Mentum* trapeziforme. *Antennæ* claviformes, articulis terminalibus transversis compressis, basalibus obconicis. *Prothorax* subcylindricus, tenuiter marginatus, antice truncatus, basi in medio paulo lobatus. *Elytra* ovata, prothoraci arcte applicata. *Pedes* breves: *femora* antica subtus unidentata: *tibiæ* curvatæ, breviter calcaratæ; *tarsi* lineares, art. ult. cæteris fere æquali. **Prosternum** postice productum, in incisura mesosterni receptum. **Metasternum** breve. *Corpus* læve, ovatum, convexum.

The form of the antennæ requires this genus to be placed in M. Lacordaire's third "group" of Helopinæ, with *Misolampus*, *Zophius*, and others, to none of which does it bear any decided affinity.   In habit it approaches some of the more convex forms of *Helops*.   The internal maxillary lobe has a very distinct hook, a character which occurs only in this subfamily in *Amphidora* and *Enoplopus*; in the latter the profemora are also toothed.   I owe my specimens to Mr. Adams, but I am unable to say from part of the coast they were derived.

Gnesis helopioides.

G. nigra, nitida; elytris fuscis, striato-punctatis.
Hab. Mantchuria.

Black, shining; antennæ, lip, and palpi ferruginous, the former shorter than the prothorax; head and prothorax finely punctured; scutellum small, acutely triangular; elytra dark brown, inclining to chestnut, deeply punctate-striate, the sutural and exterior striæ alone reaching the apex, the others uniting at various distances from it; body beneath and legs dark chestnut, shining.   Length 3½–4 lines.

Atryphodes* Macleayi.

A. niger, opacus; prothorace utrinque valde foliaceo, reflexo; elytris planatis, lineato-striatis, lateribus reflexo-marginatis, humeris distinctis.
Hab. New South Wales?

Opake, black; head with a stirrup-shaped impression in front; prothorax very deeply emarginate at the apex, each side with a broad reflexed foliaceous margin; elytra nearly perfectly flat above, the margins on each side slightly reflexed, the shoulders produced but rounded, the disk punctate-striate, the striæ more or less united anteriorly and posteriorly, and here and there single striæ within the double striæ, epipleuræ very broad and smooth; abdomen and legs glossy black.   Length 9 lines.

Atryphodes egerius.   (Pl. XIX. fig. 4.)

A. niger, opacus; prothorace antice foliaceo, postice constricto, angulis posticis acute productis; elytris obovatis, striatis, interstitiis alternis elevatis, humeris nullis.
Hab. New South Wales?

Opake, black; head with an obscure horseshoe-shaped impression in front; prothorax moderately emarginate at the apex, the sides in front rounded and slightly foliaceous, but considerably narrowed be-

* = *Thoracophorus*. Hope, not Motschulsky.

hind, each basal angle ending in a slender process directed backwards; elytra obovate, the margins very narrow, lateral, the disk striated, the striæ obsoletely punctured, the alternate intervals elevated into a sharp line or ridge; body beneath smooth, black; legs glossy black. Length 8 lines.

This and the above I received some time ago from W. Macleay, Esq., without a precise locality. They are exceedingly well-marked species, the first on account of its flat upper surface with reflected margins to the prothorax and elytra, the second from the absence of humeral angles and the remarkable slenderness of the body above the junction of the elytra and prothorax. In the last species the lateral groove on the prothorax, which on *A. Walckenaeri* is situated at the base, is placed midway between the apex and base; in *A. Macleayi* it extends to the middle, but is rather obscurely defined posteriorly. It would perhaps be more correct to regard the epipleuræ in *A. Macleayi* as limited to a portion of the under surface of the expanded margin of the elytra exterior to a slightly raised line which may be partially traced, instead of comprehending the whole: it would be more consistent with analogy, although in this case the epipleuræ would be all but confounded with the elytra.

Atryphodes errans.

A. niger, nitidus; prothorace modice convexo, glabro, marginibus angustis; elytris prothorace angustioribus, striatis.
Hab. Queensland.

Black, shining; head with a few foveæ in front (as well as the sulcations common to the genus); prothorax moderately convex, rounded at the sides, the margins narrow, latero-basal groove very short; elytra narrower than the prothorax, slightly convex, the shoulders obsolete, the disk regularly striated; body beneath and legs glossy black: the tarsi covered with golden-brown hairs. Length 7 lines.

Similar to *A. Walckenaeri*, but glossy, not opake, and a narrow margin to the prothorax, which is also rather broader than the elytra.

Atryphodes aratus.

A. niger, indumento fuscescente tectus; prothorace utrinque foliaceo; elytris subplanatis, inæqualiter striatis, lateribus reflexo-marginatis, humeris distinctis.
Hab. Queensland.

Closely allied to *A. Macleayi*, but covered with a brownish filmy crust, the elytra less flattened, and the striæ irregular, the intervening lines being somewhat flexuous, the fifth from the suture more elevated than the rest, and the two or three outer broken up into numerous oblong or roundish tubercles. The other characters are nearly the same in both species. Length 9 lines.

Adelium augurale.*

A. viridi-æneum; prothorace utrinque rotundato, modice emarginato, vermiculato-punctato; elytris interrupte costatis, punctis numerosis impressis.

Hab. Queensland.

Dark greenish bronze, in other respects resembling *A. porcatum*, but the punctures on the head and prothorax deeper and more crowded, on the latter especially running together and leaving vermicular spaces in the intervals; on the elytra the alternate ridges are considerably less elevated, or, rather, are hardly to be recognized as ridges, being more or less tubercular, the punctures more numerous and less confined to the striæ, the general surface also most minutely granulate, but without the small glabrous points visible in *A. porcatum*; body beneath and legs chalybeate blue or green. Length 6 lines.

I have seen only one specimen of this species.

Adelium succisum.

A. viridi-æneum; prothorace marginibus subfoliaceis, pone medium subito descrescente; elytris ovatis, irregulariter impresso-punctatis, punctis approximatis, aliquando connexis, punctis minoribus dispersis; antennis filiformibus.

Hab. Victoria.

Dark greenish bronze; head roughly punctured; prothorax transverse, the sides slightly foliaceous, gradually broader for about two-thirds of its length, then suddenly narrowing to the base, the disk with obscure punctiform impressions principally at the sides; elytra ovate, the shoulders nearly obsolete, covered with numerous punctiform impressions, some of which unite, and having smaller punctures in the intervals; body beneath black, very glossy, the segments on each side with a V-shaped impression; legs black, shining. Length 6–8 lines.

Narrower than *A. porcatum*, which, however, is by no means its nearest ally, and well distinguished by the peculiar angularity of the sides of the prothorax, which look as if broken off posteriorly.

Adelium vicarium.

A. fusco-æneum; prothorace marginibus rotundatis, modice dilatatis; elytris anguste ovatis, seriatim impresso-punctatis; antennis apicem versus crassioribus.

Hab. Western Australia.

Dark bronze; head punctured anteriorly, the vertex impunctate;

---

* Kirby, Trans. Linn. Soc. xii. p. 420; Lacordaire, Gen. v. p. 437.

prothorax transverse, the margins rounded and moderately dilated at the sides, finely punctured, with about three fovere on each side and a stronger impression near the margin; elytra narrowly ovate, seriate-punctate, the punctures very irregular in size and approximation, but much smaller than in the preceding species, and the spaces between them less rugose; body beneath and legs glossy black; tarsi with brownish hairs; antennæ with the last four joints but one dilated and triangular, the last rounded. Length 5 lines.

This is the only species I have noticed with what may called sub-claviform antennæ; in other respects it is clearly allied to the preceding.

Adelium obesum.

A. viridi-æneum; prothorace indistincte punctato, utrinque rotundato, pone medium constricto; elytris ampliatis, prothorace multo latioribus, fortiter punctato-striatis, humeris rotundatis.
Hab. Victoria.

Dark greenish bronze; head distinctly punctured; prothorax small, transverse, rounded at the sides, the greatest breadth being behind the middle, the breadth then rapidly contracting, the disk with numerous irregular indistinct punctiform impressions; elytra much broader than the prothorax, convex, strongly punctate-striate, the shoulders rounded: body beneath dark chalybeate green; legs glossy black; third antennal joint nearly as long as the next three together. Length 7 lines.

Adelium auratum.

A. viridi-aureum; prothorace lævi, subtilissime punctato, utrinque rotundato; elytris ovatis, interrupte fortiter striatis; corpore infra nigro, nitido.
Hab. North Australia.

Rich golden-green, with coppery reflections; head with scattered minute punctures; clypeus slightly emarginate at the apex; lip and antennæ black; prothorax smooth, finely punctured, a few irregularly distributed fovere on the disk, the sides expanded and well rounded, with the margin reflexed and thickened above; scutellum triangular; elytra broadly ovate, convex, coarsely striate, but the striæ broken up into short portions, the intervals nearly impunctate; body beneath and legs black, shining. Length 9 lines.

A distinct and handsome species, resembling our *Carabus arvensis.*

Adelium striatum.

A. æneo-fuscum, nitidum; elytris impunctatis, fortiter striatis.
Hab. Queensland.

Dark brassy brown, shining; head nearly impunctate, the clypeus very short and transverse, scarcely extending beyond the insertion of the antennæ. and having a narrow process on each side directed towards

the inner portion of the eye ; prothorax impunctate, the sides expanded
and well rounded, with the edges of the margins reflexed but only
slightly thickened, the centre of the disk with two foveæ, and behind
these towards the posterior angle an irregularly curved impressed line
on each side ; elytra shortly ovate, convex, strongly striate, nine striæ
on each ; body beneath, legs, and antennæ dark glossy brown.   Length
9 lines.

Readily distinguished by its impunctate elytra, with their strongly
marked simple striæ.   The clypeus is also peculiar.

Adelium latum.

A. late ovatum, cupreo-fuscum, nitidum ; prothorace latitudine elytrorum,
marginibus haud dilatatis ; elytris foveolatis et irregulariter lineato-
punctatis.
Hab. Australia.

Rather broadly ovate, slightly convex, dark copper-brown, shining ;
head slightly punctured ; the clypeus very short and transverse, scarcely
extending beyond the insertion of the antennæ ; prothorax as broad as
the elytra, its apex scarcely narrower than the base, thinly punctured,
and pitted with several large foveæ, posterior angles not acute, the
lateral margins not dilated, consisting merely of a thickened line ;
elytra shortly ovate, having a few irregular lines of small punctures,
here and there displaced by a broad fovea ; body beneath and legs dark
glossy brown.   Length 5 lines.

Remarkable for the breadth of the prothorax, and its moderately
convex form.   It should stand after *A. licinoides*, Kirby.

Adelium congestum.

A. elongatum, æneo-fuscum ; prothorace nitido, disperse et irregulariter
punctato, marginibus haud dilatatis ; elytris seriatim punctatis, interstitiis alternis interrupte lineato-elevatis, lineis nitidis.
Hab. Victoria.

Elongate, dark brassy brown ; head rugosely punctured ; the clypeus
very short and considerably thickened in front ; prothorax shining,
irregularly and rather coarsely punctured, not dilated at the sides, the
base strongly incurved ; elytra oblong, a little broader than the prothorax, scarcely or only very slightly shining, seriate-punctate, the
intervals between the alternate series with interrupted elevated shining
lines, or, in other words, there are five pairs of punctured lines and
four lines of short oblong ridges on each elytron ; body beneath and
legs dark glossy black.   Length 4½–6 lines.

The sculpture of the elytra is sufficiently characteristic of this very
distinct species.

The three following genera comprise the species which are excluded from *Adelium* as defined by M. Lacordaire, and which he has indicated as types of new ones, without, however, giving their characters.

OTRINTUS.

Characteres ut in *Adelio*; sed *prothorax* basi sinuatus. *Elytra* convexa, angustata, epipleuris verticalibus. *Tibiæ* anticæ curvatæ. *Mesosternum* elevatum, cavitate A-formi pro receptione processus prosterni.

Type, *Otrintus Behrii*, Germar (*Prosodes?*).

CORIPERA.

Characteres ut in *Adelio*: sed *elytra* prothoraci arcte applicata; *epipleuris* angustatis, humeris distantibus.

Type, *Coripera deplanata*, Boisduval (*Adelium*).

PHELONEIS.

Characteres ut in *Adelio*; sed *antennæ* articulis apicalibus, ultimo excepto, transversis. *Elytra* prothoraci arcte applicata. *Tarsi* antici et intermedii articulis, ultimo excepto, plus minusve latioribus, et triangulariter transversis.

Type, *Pheloneis harpaloides*, White (*Adelium*).

To any of these it is impossible to refer *Adelium catenulatum*, Boisd. It has the elytra closely applied to the prothorax as in *Coripera* and *Pheloneis*; but the epipleuræ mounting to the shoulder as in *Adelium* cuts it off from the former, while the latter has the terminal joints of the antennæ dilated, and an entirely different habit, singularly resembling, as M. Lacordaire observes, certain *Amaras*. It may be named

SEIROTRANA.

Characteres ut in *Adelio*, sed *elytra* prothoraci arcte applicata. *Antennæ* articulo tertio duobus sequentibus breviore.

Type, *Seirotrana catenulata*, Boisduval (*Adelium*).

CYMBEBA [Helopinæ].

Caput breve, retractum; *clypeus* distinctus, ultra basin antennarum haud productus, apice rotundatus. *Labrum* angustatum, productum. *Mentum* transverse cordatum. *Palpi* ut in *Adelio*. *Antennæ* lineares, art. tertio longiore, ultimo ovato. *Prothorax* transversus, lateribus vix rotundatus, leviter marginatus, apice late sinuatus, basi in medio retractus, angulis posticis rotundatis. *Scutellum* transversum. *Elytra* basi prothorace vix latiora, apicem versus sensim angustiora, lateribus subito declivia. *Pedes* ut in *Adelio*. *Prosternum* postice productum in cavitate mesosterni receptum. *Metasternum* breve. *Processus* interfemoralis antice rotundatum. *Corpus* naviculare, supra depressum.

The outline of this insect, resembling a broad species of *Elater*, and its flattened form, will readily distinguish it from all other genera of the subfamily to which it belongs.

Cymbeba dissimilis.   (Pl. XIX. fig. 8.)

C. ænea, nitida; prothorace basi biforeolato; elytris striato-punctatis, interstitiis punctulatis.

Hab. Australia.

Dark brassy brown, shining; head finely punctured, two foveæ between the eyes; clypeus separated from the front by an arched line; prothorax finely punctured, an indistinct median impressed line with a fovea on each side at the base; elytra striate-punctate, the punctures small and rather distant, the intervals between the striæ minutely punctured, the sides, abruptly declivous, shade off at the shoulder and apex; body beneath and legs brassy; tarsi beneath and last two or three joints of the antennæ covered with a ferruginous pile. Length 6 lines.

ALYMON [Megacanthinæ].

Caput retractum; *clypeus* indistinctus. *Oculi* haud approximati. *Palpi maxillares* securiformes; *labiales* art. ult. brevi, cylindrico. *Maxillæ* lobo interno mutico. *Labium* antice rotundatum. *Mentum* trapezoidale. *Antennæ* tenuatæ, art. tertio cæteris longiore, 4, 5, 6 obconicis, 7–11 crassioribus et cylindricis. *Prothorax* transversus, antice angustior, lateribus rotundatus, apice emarginatus, basi subbisinuatus. *Elytra* latissima, convexa, arcuata, lateribus subparallela, flexuosa. *Pedes* mediocres; *femora* antica subtus dente valido instructa; *tibiæ* lineares, inermes; *tarsi* art. primo subelongato. *Antepectus* brevissimum, prosterno producto. *Mesosternum* cavitate ∧-forme. *Metasternum* incurvatum.

Allied to *Oplocheirus*, Lac.; but, besides other characters, such as the unarmed internal maxillary lobe, &c., it is radically distinguished by the arching of the elytra, which may be traced along their margins and epipleuræ, and to which the concavity of the metasternum corresponds.

Alymon prolatus.

A. cupreo-fuscus, nitidus; prothorace subtiliter, elytris striato-punctatis, his interstitiis vage punctatis.

Hab. Natal.

Dark copper-brown, shining; head and prothorax finely punctured, and having a sparse greyish pubescence, apparently very deciduous; scutellum triangular; elytra rounded at the shoulders, their breadth equal to about two-thirds of their length, striato-punctate, the punctures very close, the intervals of the striæ with numerous scattered punctures, which are rather smaller than those of the striæ, the striæ eight in number on each elytron, the 4th and 5th uniting posteriorly, the 3rd and 6th also uniting beyond the two central, the remainder disappearing shortly before the apex ; body beneath, legs, and palpi brown, with a thin greyish pile. Length 6 lines.

Amaryygmus nigritarsis.*

A. oblongo-ovatus, sericeo-viridis ; capite, prothorace, antennis tarsisque nigris, femoribus tibiisque ferrugineis ; corpore infra chalybeato.
Hab. Queensland.

Oblong-ovate ; head, prothorax, and scutellum black, nearly impunctate ; antennæ more than half the length of the body, black ; elytra bright satiny green, striato-punctate, the punctures minute, the rows widely apart, also a short inner line at the base near the scutellum ; body beneath glossy black ; femora and tibiæ reddish ferruginous ; tarsi black. Length 6 lines.

Amarygmus convexus.

A. brevis, convexus, elliptico-ovatus ; capite et prothorace nigris ; elytris cyaneis, subnitidis, seriatim punctatis ; antennis pedibusque ferrugineis.
Hab. Queensland.

Very convex, short, elliptic-ovate ; head and prothorax impunctate, black ; clypeus and lip edged with ferruginous ; scutellum triangular, black ; elytra dark blue, seriate-punctate, the punctures rather coarse, the rows at moderate intervals ; two or three punctures on each side the scutellum ; body beneath black, the last two abdominal segments glossy ; antennæ, palpi, and legs ferruginous, the former about half the length of the body. Length 4½ lines.

I have a species from Sydney, nearly as convex, but much more oblong ; these are the only two very convex species I have seen.

Amarygmus tarsalis.

A. anguste ovatus, niger, subnitidus ; prothorace brevi ; elytris cyaneis, seriatim punctatis ; antennis, tarsisque subferrugineis, pedibus et corpore subtus nigrescentibus, nitidis.
Hab. Queensland.

Narrowly ovate, moderately convex ; head and prothorax opake, black, impunctate, the latter very short, the base rather strongly produced towards the scutellum, the latter triangular, black ; elytra dark blue, slightly nitid, seriate-punctate, the punctures oblong and somewhat approximate, the rows at moderate intervals, a row of seven punctures

---

* Dalman, Anal. Ent. p. 60 ; Lacordaire, Gen. v. p. 473.

on each side, near the scutellum; body beneath glossy blackish; antennæ
yellowish ferruginous, scarcely half the length of the body; legs blackish; tarsi yellowish ferruginous. Length 3½ lines.

A small narrow species, with yellowish ferruginous tarsi.

DIETYSUS [Amarygminæ].

Caput retractum; *clypeus* truncatus. *Labrum* rotundatum. *Mandibulæ* apice integræ. *Oculi* mediocres, modice approximati. *Palpi labiales* basi approximati. *Labium* transversum, emarginatum. *Mentum* trapeziforme, basi sensim angustius, ad latera margine elevato. *Maxillæ* lobo interno mutico. *Antennæ* longiusculæ, apicem versus crassiores, art. 3 elongato, 4–10 obconicis, ultimo ovato, apice oblique truncato. *Prothorax* transversus, utrinque rotundatus et tenuiter marginatus, antice angustior, basi subbisinuatus. *Elytra* ovata, prothorace basi vix latiora. *Pedes* longiusculi; *femora* sublinearia; *tibiæ* paulo curvatæ, calcaratæ; *tarsi* lineares, subtus ciliati, postici art. basali elongato. *Prosternum* productum, in cavitate mesosterni receptum. *Corpus* elliptico-ovatum, convexum.

I have long ago received specimens of this insect from M. Deyrolle, under the name here adopted. As the mesosternum presents no concavity as in the Strongylinæ, I have placed it with the Amarygminæ, with none of whose genera, however, does it seem to have any very marked affinity. The females appear to be considerably broader than the males. There are several species, one of them is from Aru.

Dietysus confusus.

D. elliptico-ovatus, fusco-æneus, nitidus; elytris striato-punctatis.
Hab. Java.

Elliptic-ovate, dark copper-brown, very glossy, and polished; head and prothorax minutely punctured; scutellum triangular; elytra finely striato-punctate, the punctures narrow and linear, the spaces between the striæ broad and flat; body beneath and legs brown, less shining than the back; anterior tarsi with the claw-joint as long as the preceding joints together. Length 6 lines.

Spheniscus* cyaneus.

S. cyaneus, nitidus; elytris brevibus, in medio obtuse elevatis, grosse et irregulariter punctatis.
Hab. Amazons.

Entirely dark blue, shining, a little paler on the elytra; head and prothorax finely punctured, the former with a large fovea between the eyes, the latter with the sides nearly straight, and with two deep foveæ

---

* Kirby, Trans. Linn. Soc. xii. p. 421; Lacordaire, Gen. v. p. 480.

on each side, the first behind the middle, the posterior nearly at the
base and the smaller of the two ; scutellum triangular, below the level
of the elytra ; elytra obtusely elevated in the middle, irregularly covered
with coarse, deep punctures of varying size ; body beneath and legs steel-
blue, the latter darker, finely punctured ; antennæ with the 7th joint
obconic, the succeeding ones more or less transverse. Length 8 lines.

Very distinct, on account of its uniform dark-blue colour, from any
of the nineteen species described in M. J. Thomson's Monograph.

Sinopium [Strongyliinæ].

Characteres ut in *Strongylio*; sed *oculi* minores. *Antennæ* breviusculæ,
art. quatuor ultimis transversis, clavam compressam formantibus. *Tarsi*
articulis, duobus ultimis exceptis, dilatatis, art. basali quam sequens
vix longiore, art. ultimo cæteris simul sumptis æquali vel longiore.

To these characters it may be added that the three or four basal
joints of the tarsi are fringed with long hairs, and that the onychium
is very distinct. The type is *Sinopium variabile*, Walker (*Strongylium*), from Ceylon.

The two following genera of Tenebrionidæ having been suggested
by M. Lacordaire, I have here given them names.

Ageonoma [Zopherinæ].
Type, *Ageonoma diabolica.*
Nosoderma diabolicum, Le Conte, Ann. Lyc. New York, v. p. 130.

Differs from *Nosoderma* in having prothoracic canals for the reception of the antennæ, and in the mandibles being entire at the
apex, and from *Zopherus* in the labium and labial palpi being exposed, the mentum broadest at the base, contracted anteriorly, and
deeply emarginate at the apex, in the subperfoliate antennæ, with
the 11th joint nearly obsolete, the absence of the transverse groove
on the last abdominal segment, &c. Lacordaire, Gen. v. p. 92.

Zygas [Adelostominæ].
Type, *Zygas cimicoides.*
Eurychora cimicoides, Quensel in Schönherr, Syst. Ins. i. p. 137, *note*.

Differs from *Steira* in having the third joint of the antennæ longer
than the first, and from *Eurychora* and *Pogonobasis* in the base of
the elytra being close to the base of the prothorax. Lacordaire,
Gen. v. p. 98, *note*.

Sessinia [Œdemeridæ].

This genus, founded on those species of *Nacerdes* with *two* spines to their anterior tibiæ, was proposed by me in this Journal in January 1863. M. Léon Fairmaire, in the French 'Annales' for August in the same year, named the group *Ananca*, apparently not aware that it had been previously disposed of.

Othelecta [Cistelidæ].

Caput exsertum, oblongum; *clypeus* transversus, distinctus. *Labrum* magnum, hirsutum. *Mandibulæ* apice integræ. *Oculi* prominuli, obliqui. *Antennæ* filiformes, art. 3 et 4 æqualibus. *Mentum* transversum. *Labium* magnum, basi pedunculatum. *Palpi maxillares* art. ult. cultriformi, *labiales* triangulari. *Prothorax* quadratus, lateribus tenuissime marginatus, apice modice rotundatus, basi truncatus. *Scutellum* parvum, transversum. *Elytra* connexa, elongato-ovalia, humeris nullis; *epipleuræ* subverticales. *Pedes* mediocres; *femora* sensim incrassata; *tibiæ* rectæ; *tarsi* haud lamelligeri, lineares, postici art. basali elongato. *Coxæ* anticæ exsertæ, approximatæ. *Prosternum* angustissimum, postice basi latiore. *Mesosternum* antice rotundatum. *Metasternum* normale. *Processus* interfemoralis quadratus, antice paulo rotundatus. *Abdomen* segmentis quinque in utroque sexu.

There are several points of structure common to this and Solier's genus *Cylindrothorus*. The pronotum, however, is quite distinct from the flanks of the prothorax, although its separation is only marked by a very delicate line, which is continuous with the ordinary border-line and dips down at the sides, so that the flanks rise into a gradually narrowing point at each extremity. The penultimate joint of the tarsi is prolonged underneath the claw-joint, but can scarcely be called lamellate. The labium is almost membranous, and is attached to the mentum by a broad peduncle. This genus appears to form the type of a distinct subfamily.

Othelecta torrida.   (Pl. XIX. fig. 5.)

O. nigra, nitida; elytris castaneis, longe et disperse pilosis.
Hab. 'NGami.

Black, shining; head and prothorax closely punctured, the punctures small, but deep and distinct, having at the bottom of each a whitish secretion; elytra brownish chestnut, oblong-oval, rather pointed behind and much broader in the female, finely but irregularly and rather remotely punctured, and having long, dispersed black erect hairs; body beneath and legs reddish chestnut, closely punctured, especially on the femora; tarsi clothed with close-set black stiffish hairs. Length 7–8 lines.

METISTETE [Cistelidæ].

Caput pone oculos paulo constrictum, antice breve ; *clypeus* transversus. *Oculi* supra subapproximati, infra distantes. *Palpi maxillares* articulo ultimo latissime triangulari. *Prothorax* subtransversus, convexus, ad latera rotundatus et tenuiter marginatus, basi sinuatus. *Elytra* anguste obovata, humeris obsoletis. Cæteris ut in *Tanychilo.*

Tanychilus has a well-developed muzzle ; this has none at all : nevertheless, notwithstanding the other characters, there is a considerable general resemblance between the two genera. The type is *Metistete gibbicollis* (*Tanychilus*), Newman (Entom. Mag. v. p. 489). *T. cistelides* of the same author may be another species, but it is unknown to me.

HOMOTRYSIS.

Characteres ut in *Allecula* ; sed *antennæ* articulo tertio quam primus duplo longiore, quarto fere æquali. *Oculi* angustati, transversi. *Tarsi* validi, antici art. basali obconico, duobus sequentibus haud longioribus, transversis.

As *Allecula* stands at present it is far from being a homogeneous genus; but taking *A. morio* as the type, *Homotrysis* differs in the form of the eyes, the proportionate length of the joints of the antennæ, and the shorter and stouter tarsi, and especially of the shorter basal joints. The type is *Homotrysis tristis* (*Allecula*), Germar (Linn. Ent. iii. p. 201). The following species, remarkable for its small prothorax, has, however, more slender tarsi, which so far weakens the force of this character.

Homotrysis microderes.

H. nigra, pilosula ; prothorace parvo, confertim punctato ; elytris fulvobrunneis, striato-punctatis ; tibiis brunnescentibus.
Hab. Victoria.

Slenderer than *H. tristis,* the elytra being three times the length of their breadth, and the head and prothorax very much narrower, and the latter also shorter : these are similarly punctured, but less pilose : scutellum black, triangular ; elytra striato-punctate, the interstices also punctured, pale fulvous-brown, except the suture, which is tinged with black ; body beneath, legs, and antennæ black and shining ; the tibiæ, except at the apex, fulvous-brown. Length 6 lines.

HYBRENIA [Cistelidæ].

Homotrysi affinis ; sed *oculi* magni, approximati. *Prothorax* basi elytrorum arcte applicatus.

The large approximate eyes will also distinguish this genus, *inter*

alia, from *Isomira*. The four anterior tarsi have all their joints
furnished with lamellæ, except of course the claw-joint; the posterior
have only the penultimate joint lamellate; but these are of scarcely
sufficient importance as generic characters.

Hybrenia insularis.

H. brunneo-ferruginea, pube grisea sparsa; elytris striato-punctatis, unicoloribus; antennis fuscis.

Hab. Lizard Island (Northern Australia).

Brownish ferruginous; the elytra paler, clothed with short sparse
semidecumbent grey hairs; head scarcely longer than broad, the lip
produced; antennæ dark brown, slender, about two-thirds the length
of the body; prothorax transverse, truncate at the apex and base,
rounded at the sides, and finely margined, narrower anteriorly, the
posterior angle not produced; scutellum triangular; elytra oblong-
ovate, the broadest part behind the middle, not broader at the base
than the base of the elytra, striate-punctate, the intervals between the
striæ somewhat transversely punctured; body beneath reddish chestnut,
finely punctured; legs brownish, with grey hairs. Length 6 lines.

Hybrenia vittata.

H. rufo-brunnea, pube grisea sparsa; elytris striato-punctatis, sutura
  vittisque duabus viridescentibus; antennis rufescentibus, fusco annulatis.

Hab. Port Albany (Northern Australia).

Light reddish brown, with a sparse greyish pubescence; head and
prothorax as in the last; antennæ pale reddish or tawny, all the joints
except the first two, dark brown at the apex; elytra narrower than in
the last, striate-punctate, the intervals between the striæ somewhat
transversely punctured, the suture and two dark-greenish stripes on
each, neither of them extending to the apex, and the antennæ commencing at some distance from the shoulder; body beneath brownish
chestnut, shining; legs tawny; femora and tibiæ with a large dark-
brown blotch on each, except on the posterior tibiæ. Length 6 lines.

Chromomæa [Cistelidæ].

Caput antice subproductum. *Labrum* rotundatum. *Mandibulæ* elongatæ.
Oculi mediocres, distantes. *Palpi maxillares* elongati, art. ult. cultriformi, *labialium* breviter triangulari. *Antennæ* breviusculæ, art. primo
gracili, tertio elongato, cæteris obconicis. *Prothorax* oblongus, fere
parallelus, ad latera tenuiter marginatus, basi truncatus. *Elytra* prothorace latiora, oblongo-ovata. *Pedes* mediocres; *tibiæ* rectæ, valide
calcaratæ; *tarsi* antici et intermedii paulo dilatati, art. ult. duobus
præcedentibus longiore, postici graciles, art. ult. elongato. *Prosternum*
elevatum. *Mesosternum* declive. *Processus* interfemoralis anguste
triangularis.

M. Bohemann's genus *Euomma** is, I have no doubt, identical
with my *Apellatus* (*ante*, p. 45), notwithstanding the author approximating it to *Eutrapela*, entirely overlooking its pectinated
claws, and the form of the last antennal joint. *Euomma* has, however, been long used for a genus of Curculionidæ ; but as we have
both happened to fix on the same name for the two species we have
respectively described, it becomes necessary to change mine as the
later of the two. I propose, therefore, to call it *Apellatus amænus*.
M. Bohemann's retaining that of "*lateralis*." *Chromomæa* has much
the same general appearance, but has small eyes, widely apart, and
a differently formed prothorax ; in the latter respect it agrees with
Æthyssius†, but which, *inter alia*, has a truncate interfemoral process. The Cistelidæ appear to be well represented in Australia.
although few species are published.

Chromomæa picta.

C. pallide-flava, subnitida, capite, prothorace, lateribus et regione suturali
elytrorum nigris.
Hab. Queensland.

Pale yellowish, slightly shining ; head and prothorax brownish black,
closely and finely punctured ; lip, antennæ, and palpi yellowish ; mandibles black ; scutellum black, very transverse ; elytra striate-punctate.
the punctures approximate, the intervals between the striæ rather broad :
sutural region, sides, and epipleuræ black ; body beneath glossy black :
legs yellowish, the apical half of the posterior thighs black. Length
5 lines.

ICTISTYGNA [Lagriidæ].

Caput subquadratum, ad angulum posticum rotundatum ; collum angusta-
tatum. *Labium* transversum. *Oculi* rotundati. *Mentum* transversum.
Labrum membranaceum, rotundatum, basi pedunculatum. *Maxillæ*
lobo externo minuto, triangulari. *Palpi labiales* parvi, apicem versus
incrassati, art. ult. triangulari ; *palpi maxillares* art. ult. securiformi.
Antennæ modice elongatæ, art. basali elongato, incrassato, 2°, 3° brevioribus, cæteris tertio longioribus, subobconicis, ultimo præcedente longiore.
Prothorax late ovatus, antice in collum constrictus, lateribus rotundatus, postice angustior. *Elytra* oblonga, angustata. *Tibiæ* bicalcaratæ.
anticæ extus spinosæ ; *tarsi* art. primo elongato, penultimo subbilobo.
Mesosternum angustissimum. *Corpus* elongatum, hirsutum.

This genus and *Diacalla* (*ante*, p. 46), should, I think, form a
separate subfamily among the Lagriidæ, distinguished by their
rounded eyes, spurred tibiæ, and peculiar habit, from the more

* Fregat. Eugenies, Ins. p. 101, pl. 2. fig. 1.
† *Atractus* (MacLeay), Lacord. *nec* Laporte (see *ante*, p. 45, *note*).

typical *Lagriæ*.   *Diacalla* differs from this genus, *inter alia*, in the
form of the head and prothorax, in the shorter antennæ, in the
greater length and cultriform shape of the external maxillary lobe,
in the longer and cylindrical labial palpi, &c. &c.   The spines on the
anterior tibiæ are almost hidden by the pubescence.

Ictistygna vetula.

I. fusca, elytris pedibusque rufulis.
Hab. New South Wales.

Dark brown, with a thin greyish pubescence, mixed with long erect
hairs; head and prothorax covered with coarse crowded punctures;
scutellum quadrate, with a close silvery pubescence; elytra nearly
twice the breadth of the prothorax at the base, dull reddish, shining,
coarsely punctate, the punctures deeply impressed and nearly contiguous, with the intervals very rugose; body beneath reddish brown
or dark brown, with a greyish pubescence; legs reddish, pubescent;
antennæ about half the length of the body, the second joint a little
shorter than the third.   Length 5 lines.

Ictistygna adusta.

I. nigra, elytris pedibusque rufulis, tarsis anticis quam in præcedente longioribus.
Hab. New South Wales.

Smaller, and more slender than the former, with the head, prothorax,
and antennæ black, the second joint of the antennæ much shorter than
the third, the anterior tarsi longer and narrower, and the anterior tibiæ
less spined on their outer edge.   Length 2¾ lines.

Eucioides [Anthribidæ].

Caput infra oculos paulo latius; *rostrum* transversum, antice truncatum;
scrobe magna, rotundata.   *Antennæ* maris corpore plus duplo longiores,
art. primo brevi crasso, secundo elongato-obconico, reliquis ad nonum
capillaribus, apice nodosis, duobus ultimis brevibus clavam formantibus.
Oculi mediocres, laterales, armati.   *Prothorax* subcylindricus; *carina*
basi haud parallela, et ad latera vix continuata.   *Elytra* cylindrica,
prothorace paulo latiora.   *Pedes* mediocres; *tarsi* art. basali sequentibus duobus simul sumptis haud longiore.

Most of the characters of this genus are those of *Exillis**, after
which it may be ranked.   It has, however, the head broader below,
so as to form a strong margin round the scrobe, the second antennal
joint elongate and obconic, the ninth entering less decidedly into

* Pascoe, Ann. and Mag. Nat. Hist. ser. 3. v. p. 43.   *See* also Lacordaire
Genera, &c. vii. p. 583.

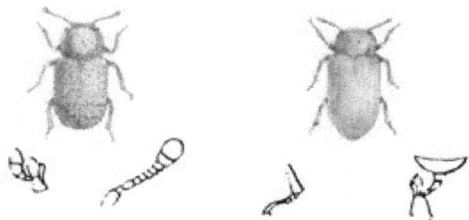

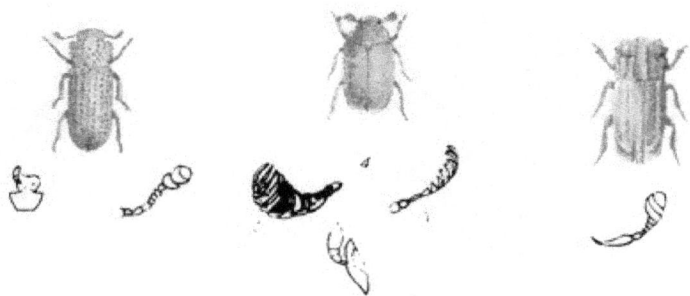

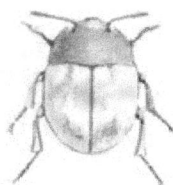

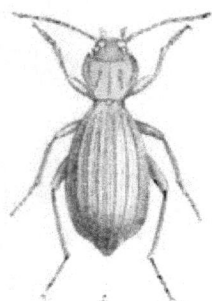

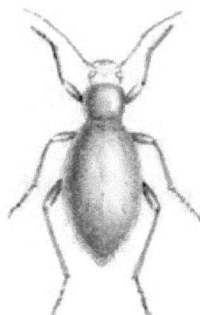

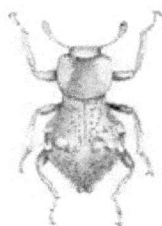

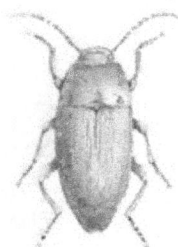

the club, and the club itself much shorter, the carina of the prothorax only very slightly prolonged at the sides, and the basal joint
of the tarsi much shorter. The female is stouter and larger, with
antennæ scarcely longer than the body. The amount of pubescence
varies according to the individual. For this interesting addition to
the scanty list of Australian Anthribidæ I am indebted to Mr.
Odewahn.

Euciodes suturalis.

E. nigra, pilis albis sparse induta, præcipue in regione suturali.
Hab. South Australia.

Black, with a slight brassy tinge, sparsely and irregularly covered
with longish chalky-white hairs; antennæ black, the two basal joints
sometimes reddish testaceous; eyes black, forming a regular arch above
the scrobe; prothorax rounded, and a little narrower anteriorly; scutellum oblong; elytra punctato-striate, but the striæ nearly concealed
by the pubescence; body beneath with a close white pile; legs black;
the tibiæ reddish testaceous. Length 1¼-2 lines.

EXPLANATION OF THE PLATES.

[*From the* Annals and Magazine of Natural History *for*
January 1869.]

DESCRIPTIONS

OF

NEW GENERA AND SPECIES

OF

TENEBRIONIDÆ

FROM

AUSTRALIA AND TASMANIA.

BY

FRANCIS P. PASCOE, F.L.S., F.Z.S., &c.

[Plate X.]

DR. HOWITT, of Melbourne, having recently sent me a large
collection of Heteromera from Australasia and New Zealand, I
propose to describe in this Magazine such of the new Australian
species as belong to the family Tenebrionidæ, adding several
more derived from other sources, leaving the remainder and
those from New Zealand for a future opportunity.

The Tenebrionidæ* belong preeminently to the hot and dry
regions of the earth; the epigeous or more normal forms are
found in very small numbers, either in the humid lands of the
tropics or in the northern parts of the northern hemisphere.

---

* In the sense in which it is constituted by M. Lacordaire (Gen. des
Coléopt. t. v.). The great advantage of having a standard which is in
everybody's hands appears to me to render it desirable to conform as
closely as possible to the classification and to the principles of analysis
applied to the characters of the various divisions of the family. Only, for
the sake of greater simplicity, I have called his " tribus " and " groupes "
(the latter often of equal rank with the former) subfamilies. The " sec-
tions " and " cohorts," being merely designations of the primary branches
of a dichotomous arrangement, do not themselves form natural divisions.

A

England contains only seventeen (or, with the doubtful and introduced, twenty-seven) species, while the countries surrounding the Mediterranean have, according to M. de Marseul's Catalogue, 1327 species. From Australia and Tasmania we have about 210 described—a number probably far below that contained in the rich collections of Melbourne and Sydney, and which we cannot doubt will be still greatly increased as those countries are more explored. The lists which Dr. Howitt has favoured me with from time to time bear evidence of the narrow limits in which a large number of species are localized.

There is some confusion in regard to the use of the terms for those parts of the elytra known as the "epipleura" and the "epipleural fold"*, which it is necessary to notice: when only one is present or strongly marked, either term is often used indifferently; while the former, in a second sense, is supposed to express generally the descending or inflected sides of the elytra. In future I propose to use the term "epipleura" for that part of the flank of the elytron marked off from the rest by a line more or less sharply defined; when there is a descending side above this line, as in *Zopherosis*, I propose to call it the "pleura." This should have been the epipleura, if the word had been used in the strictest sense; but it is too late now to attempt to alter its ordinary signification. The stripe along the lower border of the epipleura will be the "epipleural fold" (*plica epipleuralis*); when nearly obsolete, there is still very often a sort of raised line or border which marks its position. Good examples of well-marked epipleura and epipleural fold, without the pleura, will be found in our common *Blaps mortisaga*, or, still better, in the genus *Acis* (*Akis*).

ORCOPAGIA.

Subfamily *BOLETOPHAGINÆ*.

Antennæ clavatæ, 10-articulatæ; clava biarticulata.
Tibiæ anticæ crescentiformes.

Head vertical, deeply sunk in the prothorax, excavated in front between the eyes and clypeus, the latter cornuted, the lip lying in the space between the mandibles; antennary ridge bilobed. Eyes small, transverse, impinged on by the antennary ridges, but not divided. Antennæ clavate, ten-jointed, the scape elongate; the third joint longer than the second, the rest to the eighth gradually shorter, the ninth and tenth forming a large oval pubescent club, the latter twice as large as the

---

* "*Repli épipleural*" of M. Lacordaire. "Fold" is a bad rendering of "*repli*," but I know of none better. Dr. Leconte does not appear to notice this part.

former. Mentum subcordiform; lower lip transverse, broadly emarginate, and fringed anteriorly, its palpi short, with the last joint large, obovate. Maxillæ with the lobes of equal breadth; the palpi moderate, with the last joint cylindrical and obliquely truncate. Prothorax transverse, rounded, crenate, and expanded at the sides, but not foliaceous, elevated and compressed above, and projecting over the head at the apex. Elytra elongate, parallel, narrower than the prothorax, posteriorly abruptly declivous, sides nearly vertical; the epipleuræ indistinct. Legs short; femora not thickened; tibiæ compressed, the outer edges 5–6-toothed, the anterior crescent-shaped, the intermediate arched externally. Prosternum elevated, rounded, not produced behind. Mesosternum entire. Metasternum moderately elongate. Intercoxal process narrowly triangular. Body tuberculate; prothorax and elytra above in an even plane throughout.

There are three genera of Boletophaginæ with ten-jointed antennæ: one is North American (*Phellidius*[*], Leconte), another (*Ozolais*, Pasc.) is from Ega, on the Amazons, and the above[†]; as might be expected from three such widely separated localities, there is very little affinity between them. There are several genera, some new, with eleven-jointed antennæ, which, as they do not belong to Australia, I propose to consider in a future article: one of them has been recently published as a *Diceroderes* (*D. elongatus*, Redtenbacher), but it is a true Boletophagin (*Dysantes*, MS.).

Orcopagia monstrosa. Pl. X. fig. 8.

O. elongata, indumento rufo-ferrugineo vestita, subtus pedibusque squamosis.

Hab. Clarence River.

Elongate, covered above and on the head with a reddish-ferruginous felt-like substance; beneath and legs with small scales of a yellower colour; head completely concealed above by the prothorax, the horn on the clypeus horizontal (in reference to the body); prothorax longitudinally excavated above, the excavation bordered above with a row of tubercles, except posteriorly, where it is also notched for the reception of part of the scutellum; the latter oblong rounded, a little raised; elytra irregularly tuberculate, particularly a strongly marked crest, which is also tuberculate, on each side of the scutellum, and projecting forwards on the prothorax at the edge on the

---

[*] = *Boletotherus*, Candèze. The name in the text has priority.

[†] It was briefly characterized by me in the Proc. Entom. Soc. for April last (1868).

declivous portion on each side a conical tuberculate projection.
Length 4 lines.

ULODICA.

Subfamily *ULODINÆ*.

Antennæ haud clavatæ; art. 3⁽ᶦᵒ⁾ quam 4⁽ᵗᵘˢ⁾ duplo longiore.
Prothorax transversus, utrinque rotundatus, marginibus squamosis.

This genus differs from *Ulodes** in its antennæ having
the third joint much longer than either the second or fourth.
Ulodes has the remarkable character of having all the joints of
equal length, the last three, as in *Ulodica*, being pubescent,
while all the others are covered with stiff scale-like hairs
arranged in dense whorls. The genus was referred by its
author, as well as by M. Lacordaire (to whom, however, it was
unknown), to the vicinity of *Boletophagus*. From the subfamily to which the latter belongs, all the species, as well as
those of the cognate genera which have come under my notice,
differ in being destitute of the transverse excavation which occurs behind the insertion of the mentum of the *Boletophaginæ*;
and, so far as I know, they have globose, not cylindrical, anterior coxæ. Probably, if the illustrious author of the 'Genera'
had known any of the species, he would have made *Ulodes*
the type of another group, as I have now ventured to do.
The four genera which constitute the subfamily at present
may be tabulated thus :—

Antennæ clavate .... *Ganyme*, Pasc.
Antennæ not clavate.
  Prothorax scaly at the sides.
    Antennæ with the third joint longest .... *Ulodica*, Pasc.
    Antennæ with the third joint not longer than the
    rest .... *Ulodes*, Er.
  Prothorax ciliated at the sides .... *Dipsaconia*, Pasc.

Ulodica hispida.

U. oblonga, fusca, dense brunneo-nigroque squamosa; prothorace
disco quadri-verrucoso-fasciculato.

Hab. Clarence River.

Oblong, dark brown, closely covered with pale reddish brown,
varied with black, scales; head with small dull reddish-brown
scales; antennæ brownish grey ringed with black—principally
the third and fourth, sixth, eighth, and base of the ninth joints;
prothorax roughly scaly, the apex with two wart-like tubercles
clothed with a bunch of erect blackish scales; behind the mid-

* Erichson in Wiegmann's Archiv, 1842, i. p. 180, Taf. 5. fig. 1. To this
genus also belongs *Endophlœus variicornis*, Hope; the same author's *E.
australis* is a *Dipsaconia*.

dle two similar tubercles, but of a pale brownish colour, like the rest of the disk, except a small black spot on the margin on each side; scutellum transversely oblong, scaly; elytra striato-punctate, the alternate interstices with small, blackish, wart-like tubercles, which are obscured by irregular black patches, giving the elytra a dull brownish ferruginous hue; body beneath and legs ferruginous, with greyish-yellow scales; tibiæ with a black ring in the middle. Length 3½ lines.

Dr. Howitt has also sent me a specimen of this species, but without a locality.

GANYME.

Subfamily ULODINÆ.

Antennæ clavatæ, art. 3[io] quam 4[tus] longiore.
Oculi transversi, angustati.
Prothorax utrinque fortiter angulatus.

Head small, inserted into the prothorax nearly to the eyes, a little produced in front; clypeus indistinct; antennary ridge very small. Eyes prominent, transverse, narrow throughout. Antennæ clavate, the joints, except the last three, surrounded with whorls of stiff hairs; scape not stouter than the other joints, the third twice as long as the second, and longer than the fourth, the remainder to the eighth becoming gradually shorter, ninth and tenth transverse, eleventh rounded, discoloured, the last three forming a short pubescent club. Oral organs apparently as in *Ulodes*, but the labium less transverse and more decidedly quadrate. Prothorax short, transverse, apex strongly emarginate, each side expanding into a broad pointed angle extending from the apex to the base, and fringed with short, curved, stoutish hairs; the base broadly lobed; the disk slightly convex, irregular. Elytra rather short, much broader than the prothorax, convex, slightly irregular, not costate, broadest at the base, the shoulders rounded and prominent. Legs shortish; tarsi slender, slightly hairy beneath, the posterior claw-joint not so long as the rest together; anterior coxæ globose, not approximate. Prosternum flat. Metasternum moderately long.

A well-marked genus, on account of its peculiar prothorax and clavate antennæ. In colour and clothing the species described below bears a striking resemblance to *Lemodes coccinea*, Boh., an anomalous form supposed to belong to the Pyrochroidae, common in fungi under the bark of decaying trees in Victoria. *Boletophagus Sapphira*, Newm.[*], is another member of this genus, larger and more brightly coloured, with the suture and borders of the elytra black.

* Entom. i. p. 104.

Ganyme Howittii.   Pl. X. fig. 7.

G. sordide miniacea, subsericea; antennis, art. ultimo excepto, pedibusque nigris.

Hab. Victoria; Tasmania.

Closely covered with a dark miniaceous, somewhat silky, scale-like pubescence, paler, less dense, and more scale-like beneath, and without a vestige of punctuation; upper lip and palpi brownish black; antennæ black, except the last joint, which is of a reddish-white colour; prothorax with two vague impressions in front and two behind, the latter more towards the sides; scutellum cordiform, indistinct; elytra short in proportion to the breadth, but about four times the length of the prothorax, very convex, irregular, rather abruptly declivous behind, one little callosity behind the shoulder, and two on the declivity, the epipleura curving sharply up towards the shoulder; legs black, the tips of the tibiæ and tarsi inclining to ferruginous.   Length 2 lines.

Melytra.

Subfamily *Apocryphinæ*.

Antennæ apice paulo incrassatæ, art. 3^io sequentibus multo longiore.

Mentum subquadratum; palpi labiales art. ultimo conico; labium membranaceum.

Maxillæ lobo exteriore brevi, transverso; palpi maxillares art. ult. subsecuriformi.

Head triangular, subvertical, inserted into the prothorax nearly as far as the eyes; antennary ridge almost obsolete. Eyes prominent, round, entire. Antennæ exposed at their insertion, long, filiform, but a little thicker at the apex; scape globose-ovate, second joint obconic, third twice as long as the scape, fourth to the eighth much shorter than the third, ninth and tenth thicker than the preceding, eleventh elongate-ovate. Mentum subquadrate; lower lip very small, membranous. Maxillæ very short; outer lobe transverse, inner unarmed. Maxillary palpi long, with the last joint securiform; last joint of the labial palpi conic.   Prothorax oblong, a little depressed, slightly rounded at the sides, the flanks confounded with the pronotum, base and apex truncate. Elytra rather short, ovate; epipleura vertical, narrow, with the flanks of the elytra raised above them, the shoulders obsolete; no wings.   Legs moderate; femora thickened; tibiæ filiform; tarsi narrow, all nearly equal, the claw-joint elongate.   Anterior coxæ globose, exserted, not approximate.   Prosternum on the same plane with the rest of the propectus; the anterior cotyloid cavities rather remote from its posterior edge, intermediate with trochantins

angulated externally. Metasternum shorter than the mesosternum. Interfemoral process rather narrow, triangular.
Abdomen with the ventral segments nearly equal in length.

This genus and the following are so far connected that in
both the flanks of the prothorax are not separated from the
pronotum, and the mentum is sessile to the throat. In other
respects their principal characters are very dissimilar. For
further remarks I must refer to the next genus.

Melytra ovata. Pl. X. fig. 1.

M. subnitida; capite et prothorace nigro-piceis; elytris cupreis;
antennis pedibusque ferrugineis.

Hab. Tasmania.

Subnitid; head and prothorax pitchy black, finely punctured; palpi and antennæ light ferruginous, the latter more
than half the length of the body, and paler at the apex; scutellum transversely triangular, acuminate behind; elytra copper-brown, seriate-punctate, the punctures rather coarse and
somewhat longitudinally impressed, the intervals between the
rows minutely punctate; body beneath chestnut-brown, finely
punctate; legs light ferruginous. Length 3 lines.

HYMÆA.

Subfamily *APOCRYPHINÆ.*

Antennæ clavatæ, art. tertio sequentibus haud longiore.
Mentum transversum, antice gradatim angustius; labium corneum.
Maxillæ lobo exteriore elongato, angustato; palpi maxillares art.
   ultimo ovato.

Head subtriangular, rounded and obtuse anteriorly, subvertical, inserted into the prothorax nearly as far as the eyes;
the clypeus separated from the front by a deep slightly arched
suture; antennary ridge small, auriform. Eyes conically projecting, round, entire. Antennæ exposed at their insertion,
subelongate; scape globose, second joint shortly turbinate,
third to the eighth elongate-turbinate, nearly equal in length,
ninth and tenth nearly equilaterally triangular, eleventh ovate,
pointed, not longer than the tenth, the three forming a depressed
club. Mentum transverse, rounded at the sides, gradually and
rapidly narrowing towards the insertion of the lower lip, the
latter small, rounded, corneous. Maxillæ narrow, the inner
lobe unarmed. Maxillary palpi long, with the last joint ovate,
of the labial shortly cylindrical. Prothorax oblong, narrowed
posteriorly, the sides rounded, the flanks confounded with the
pronotum, apex and base truncate. Elytra short, ovate, the
shoulders obsolete; epipleura narrow, vertical; no wings.

Legs moderate; femora thickened in the middle; tibiæ gradually stouter towards the apex; tarsi lengthened, slender, the claw-joint moderate. The under parts nearly as in the preceding genus, but the anterior cotyloid cavities very close to the posterior border of the propectus, the mesosternum and metasternum a little longer, the interfemoral process very considerably broader, and the ventral segments gradually decreasing in length to the fourth.

The position of *Hymæa* and *Melytra* is somewhat doubtful. From the characters of the "*Apocryphides*," as given by M. Lacordaire[*], they seem to me to belong to them. Mr. F. Bates, who has made the Heteromera his especial study, inclines to the opinion (*in litt.*) that, from the narrow antennary ridges, they are more nearly related to the *Strongyliinæ*, and that they form a distinct subfamily. In the 'Genera,' the "*Apocryphides*" are classed among the "*Hélopides*," an arrangement to which Dr. Leconte[†] objects, because of the absence of the membranous margin of the third and fourth abdominal segments, "which is so evident in Helopini and all the allied tribes." He admits, however, that "the observation of such characters as are relied on for the classification of this family is sometimes very difficult in small species, unless specimens may be submitted for dissection." *Hymæa*, as it appears to me, has entirely corneous ventral segments, while *Melytra* has the third and fourth segments membranous posteriorly. Both have the mentum without a pedicel, and the base of the maxillæ and lower lip exposed. There are trochantins[‡], I think, in both. At any rate, their intermediate cotyloid cavities are angulated externally. M. Lacordaire ascribes trochantins to *Apocrypha*, although he says it is difficult to decide if they really exist. Dr. Leconte refuses them without any doubt. With regard to the antennary ridges, it sometimes happens that the difference between the continuous ridge (Platygene) and the narrowed and more limited ridge (Otidogene) is one of degree, leaving it doubtful to which category they belong. Dr. Leconte places his two North-American "tribes" Meracanthinæ and Strongyliinæ (both otidogenous) in his "subfamily Tenebrionidæ (genuini)" together with Blaptinæ, Boletophaginæ, Helopinæ, and many others, all platygenous—an arrangement very

---

[*] Genera, &c. v. p. 432.   [†] Classif. Col. North Am. p. 218.

[‡] The trochantin is a small piece attached to the outer edge of the coxa; in the Tenebrionidæ, when it is present, it is confined to the intermediate pair, and it is generally, if not invariably, correlated with a cotyloid cavity having a very pronounced angle over the spot where it occurs. I have given a diagram of the coxa with a trochantin attached on Pl. X. fig. 9.

different from M. Lacordaire's, and attaching to the character a much less degree of importance than is done by him. The strongest argument against placing *Hymæa* and *Melytra* among the *Apocryphinæ* is that the mentum is attached to the throat without the intervention of a pedicel.

Hymæa succinifera. Pl. X. fig. 3.

H. nitida, fulvo-brunnea; elytris tuberculis succineo-flavis instructis.

Hab. Tasmania.

Shining fulvous brown; head rather coarsely punctured; prothorax not broader than the head measured across the eyes, coarsely punctured, the intervals here and there raised into small tubercles; scutellum large, but its limits very indistinct; elytra scarcely longer than the head and prothorax together, seriate-punctate, the punctures large and connected by a slight longitudinal impression, a few erect, stiffish hairs scattered chiefly at the sides; on each elytron towards the outer side two rows of large, oblong, amber-like tubercles, the outer of them of three (one on the shoulder), the inner of two tubercles, and one or two spots of the same amber-colour; body beneath brownish ferruginous, coarsely punctured; antennæ and legs yellowish ferruginous, with a few longish scattered hairs. Length 2 lines.

Atryphodes Howittii.

A. viridi-æneus, aureo-versicolor, nitidus; prothorace transverso, angulis anticis rotundatis, lateribus modice foliaceis, rotundatis, sulcis discoidalibus leviter impressis; elytris costis alternis minoribus.

Hab. Kiama.

Greenish bronze, with varying golden reflections, shining; antennæ pitchy black; prothorax transverse, broader than the elytra, anterior angles rounded, the sides with a moderately wide foliaceous margin, slightly rounded, narrower at the base, the discoidal lines shallow, the lateral abbreviated; scutellum subcordiform; elytra about twice the length of the prothorax, their alternate costæ much smaller than the others; body beneath and legs pitchy brown, shining. Length 10-11 lines.

Atryphodes is perhaps better known under its old name *Thoracophorus* [*]; but, as that name had been previously used

---

[*] Erichson said long ago, "The name must be altered, not only because it has been already used, but also because it does not comply with the rules of nomenclature." Wiegmann's Arch. 1842, ii. p. 239. *Thoracophorus,* however, in Motschulsky's sense, has been adopted by Dr. Gemminger and Baron von Harold in their great 'Catalogus Coleopterorum,' now in course of publication.

by Motschulsky, I propose l to replace it by the above*.
The characters as given by M. Lacordaire† apply to all the
species hitherto described, and therefore they need not be repeated here.   Only one species was then known (*A. Walcknaeri*, Hope); the other two, *dilaticollis*, Guér., and *Kirbyi*,
Sol., I have no doubt are referable to it. The above is a very
handsome species, and easily distinguished by its colour. All
the species appear to have the head and prothorax impunctate,
or nearly so, the former has a frontal horseshoe-shaped or
stirrup-like impressed line, the anterior portion being the
groove dividing the clypeus from the front; on the prothorax
there are a central and two lateral impressed lines, each terminating posteriorly in a more or less strongly marked fovea;
the lateral lines are frequently abbreviated. The males have
the anterior tarsi slightly dilated, and the antennæ thicker
than in the females.   I am not sure that the greater breadth
of the prothorax noticeable in some individuals is always a
sexual character.

Atryphodes Castelnaudi.

A. niger, vix nitidus; prothorace transverso, angulis anticis obtusis,
lateribus rotundatis, modice foliaceis, sulcis discoidalibus subtiliter
impressis; elytris subnitidis, costis alternis minoribus.

Hab. Kiama.

Black, scarcely or only very slightly nitid on the head and
prothorax, more so on the elytra; antennæ nitid, especially at
the base; prothorax transverse, not broader than the elytra,
anterior angles obtuse, the sides with a moderately wide foliaceous margin, well rounded, and considerably narrower at the
base; the discoidal lines nearly obsolete, except at the base,
the foveæ in which they terminate very shallow; scutellum
subcordiform; elytra about twice the length of the prothorax,
their alternate costæ smaller than the others; body beneath
and legs pitchy black, shining.   Length 10–11 lines.

I have dedicated this fine species to Count F. de Castelnau,
who, in addition to numerous previously well-known entomological works, has recently presented us with an apparently exhaustive list of the Australian Cicindelidæ and Ca-
rabidæ.

Atryphodes cordicollis.

A. niger, nitidus; prothorace subcordiformi, lateribus modice foliaceis, antice fortiter rotundatis, postice conniventibus, angulis anticis late rotundatis, sulcis discoidalibus fortiter impressis, lateralibus elongatis; elytris costis æqualibus.

Hab. Brisbane.

Black, shining; included part of the stirrup-shaped impres-

sion of the front raised above the surrounding parts; prothorax somewhat heart-shaped, the sides with a moderately
wide foliaceous margin, strongly rounded anteriorly, gradually
contracting behind into a narrow base; anterior angles broadly
rounded; discoidal lines strongly impressed, the two lateral
nearly extending to the apex, becoming, however, gradually
fainter; scutellum deeply ensconced between the elytra, rounded
posteriorly; elytra more than twice the length of the prothorax,
their costæ equal; body beneath and legs glossy brownish
chestnut, tarsi ferruginous.   Length 9–10 lines.

The strongly marked form of the prothorax is exclusively
the character of this species.

Atryphodes æricollis.

A. niger, nitidus; capite prothoraceque æreo-brunneis, hoc transverso, angulis anticis obtusis, marginibus sat late foliaceis, sulcis
discoidalibus lateralibus interruptis; elytris costis æqualibus.

Hab. Queensland.

Black, shining; head and prothorax bronze-brown, the
former with the frontal impression somewhat hexagonal, the
upper line forming three shorter sides; antennæ black;
prothorax transverse, strongly rounded and rather broadly
foliaceous at the sides, the anterior angles obtuse, lateral discoidal lines interrupted; scutellum triangular, on the same
level as the elytra; the latter about twice the length of the
prothorax, their costæ equal; body beneath and legs glossy
brownish black.   Length 6 lines.

This species in habit more nearly approximates, although
very different, to *A. Howittii*; but the strongly rounded prothorax is more characteristic of *A. Walcknaeri*. Its precise
habitat is uncertain.

Atryphodes encephalus.

A. angustatus, niger, nitidus; prothorace oblongo, antice sat fortiter
emarginatus, lateribus anguste foliaceis, modice rotundatis, sulcis
discoidalibus lateralibus interruptis vel fere obsoletis; elytris
costis æqualibus.

Hab. Rockhampton.

Narrow, black, shining; part within the frontal impression
raised and marked above with two foveæ; prothorax oblong,
sides slightly rounded, foliaceous margin of moderate width,
anteriorly rather strongly emarginate, the anterior angles
somewhat obtuse, central discoidal line well marked, the two
lateral interrupted, occasionally nearly obsolete; scutellum
triangular, lying below the level of the elytra; the latter about

the width of the prothorax and nearly twice as long, their costæ equal; body beneath and legs glossy pitchy brown. Length 7 lines.

A narrow species, readily distinguished by its strongly emarginate prothorax.

Atryphodes pithecius.

A. niger, subnitidus, elytris cupreo-fuscis; prothorace paulo convexo, utrinque modice rotundato, marginibus anguste foliaceis, sulcis lateralibus nullis.

Hab. Queensland.

Black, slightly nitid, the elytra dark copper-brown; antennæ brownish, much more slender in the female; prothorax rather longer than broad, slightly convex, the anterior angles obtuse, the margins narrowly foliaceous, the sides most rounded anteriorly, straighter behind the middle, not incurved at the base towards the posterior angle, which is therefore obtuse, the lateral dorsal grooves represented only by the foveæ at the base; scutellum small; elytra as broad as or broader than the prothorax, ovate, the costæ equal in breadth; body beneath and legs glossy brown; tarsi ferruginous.   Length 7–8 lines.

Allied to *A. errans*, Pasc., a black glossy species, but differing essentially, *inter alia*, in the form of the prothorax, which is longer, considerably less rounded posteriorly, and with the foveæ, but without any trace of the lateral grooves.   I have four specimens, all slightly differing, *inter se*, but agreeing in the characters given above. Another very near may hereafter, on more extensive examination of specimens, be found distinct.

The species of *Atryphodes* form three divisions: all above described, together with *errans* and *brevicollis**, belong to the *Walcknaeri* category, and are more or less glossy, with the foliaceous margins of the prothorax below the general level of its disk; the second category contains *Macleayi*, *aratus*, and *egerius*, and are opaque, with the margins directed upwards, especially in the two former, and the disk of the prothorax flat and lying below them; lastly, there is the following species, in which the foliaceous margins become obsolete.

Atryphodes caperatus.

A. angustatus, niger, nitidus; prothorace oblongo, angulis anticis leviter rotundatis, lateribus haud foliaceis, in medio haud rotundatis, ad basin subito contractis, sulcis discoidalibus interruptis.

Hab. Hunter's River; Darling Downs.

Narrow, black, shining; frontal space with five foveæ (three

* Redtenbacher, Novara-Reise, p. 130.   The "*licinoides*" of the same author appears to be synonymous with *aratus*.

above, two below) ; prothorax oblong, slightly broader than the elytra, sides moderately rounded anteriorly, then nearly straight, but narrowing posteriorly, near the base rounded, and then suddenly contracted and passing into the usual acute basal angle ; no foliaceous margin, the two lateral discoidal lines broken up and irregular, but varying in different individuals ; scutellum transverse, scarcely below the level of the adjacent part of the elytra ; the latter considerably more than twice the length of the prothorax, and with a bronze tint, their costæ equal ; body beneath and legs glossy brownish black, the first two abdominal segments with a more or less decided broad longitudinal depression. Length 9 lines.

A very narrow form, without foliaceous margins to the prothorax, and in these respects leading to *Otrintus.* The frontal foveæ are, in one of my specimens, connected with the upper central one by impressed lines ; in another there are four or five irregular undefined depressions.

BLEPEGENES*.

Subfamily *ADELIINÆ.*

Caput exsertum, culmen supraantennarium in spinam productum.
Maxillæ lobo interiore majore, subquadrato, apice dense fimbriato.
Prothorax apice truncatus.
Elytra costata, plica epipleurali ad humerum haud attingente.

Head exserted, gradually narrower behind the eyes, the antennary ridge prolonged into a nearly erect, slightly recurved spine ; clypeus very thick, rather suddenly bent down anteriorly, its apex emarginate, separated from the front by two fine oblique lines not meeting in the middle. Eyes transverse, narrow, entire. Antennæ filiform ; the scape obconic, the third joint not so long as the fourth and fifth together, thickened at the tip, the rest to the tenth subequal, obconic ; the eleventh not dilated, longer than the preceding joint. Mentum very narrow at the base, spreading and rounded at the sides and anteriorly ; lower lip transverse, bilobed, its palpi small. Maxillæ small, densely fringed, the inner lobe larger than the outer and unarmed ; their palpi slender, the basal joint elongate, the last securiform. Prothorax depressed, spined at the sides, apex narrowed, truncate, posterior angles obliquely

---

* This genus, with its type, was shortly described by me and published in the Proc. Ent. Soc. for April 1868. From some error, " *Clypeus valde* " was printed " *Clypeus haud.*" M. Preudhomme de Borre some time after published a description of the same species, in the 'Annales' of the Belgian Entomological Society, under the name of *Coradelium armatum.*

truncate. Elytra oblong-ovate, costate, flat above; epipleura terminating before the apex, the epipleural fold slightly sinuate, not extending to the shoulder. Legs rather long; femora and tibiæ slightly compressed; tarsi slender, the anterior in the males rather strongly dilated, the penultimate joint of all sub-bilobed. Sterna and abdomen as in *Adelium* and *Atryphodes*.

Although this genus has the subbilobed tarsi of *Adelium*, its affinity appears to me to be nearer *Atryphodes*, on account of its costate elytra, only slightly sinuate epipleural fold, and habit; in the latter respect it approaches *Atryphodes egerius*. It is among the most remarkable genera of Tenebrionidæ. The earliest specimens of this species which I saw were stated to be from Queensland; Dr. Howitt, however, gives Kiama as the habitat of the individuals he has kindly sent me.

Blepegenes aruspex. Pl. X. fig. 2.

B. cupreo-fuscus vel -niger, subopacus; elytris costis quatuor nitidis.

Hab. Kiama.

Dark copper-brown or bronze, sometimes bronze-black, nearly opaque; head and prothorax impunctate, the latter with four foveæ on the disk, or the lateral foveæ are connected and form an irregular longitudinal impression, each side before the middle expanding into a strong triangular spine, subhorizontal or directed a little upwards; near the base a much smaller spine or tooth, the posterior part of which slopes directly inwards to the base; scutellum transversely triangular; elytra more than three times the length of the prothorax, each with four glossy costæ, none of them reaching to the apex, the sutural and second costa having a less elevated opaque costa between them, each apex ending in a short diverging mucro; legs ferruginous brown, shining; body beneath very glossy, brown; antennæ ferruginous. Length 8–9 lines.

BYALLIUS.

Subfamily ADELIINÆ.

Antennæ art. tertio elongato, cylindrico.
Frons parum convexa, sulci longitudinales nulli.
Maxillæ lobo interiore unciformi.
Elytra obovata, reticulata, plica epipleurali obsoleta.

Head deeply inserted into the prothorax, the front slightly convex, without any grooves; the clypeus broadly truncate at the apex, separated from the front by a narrow, distinct, arched line. Eyes transverse, impinged on by the antennary ridges. Antennæ filiform; scape obconic, the third joint cylindrical, longer than the fourth and fifth together; the two latter and remainder to the tenth obconic, becoming very gradually shorter; the eleventh longer, ovate, depressed. Mentum rather narrow behind, rounded at the sides anteriorly; lower lip transverse, slightly emarginate and fringed at the apex, largely excavated in the middle on each side for the insertion of the labial palpi. Maxillæ with the inner lobe narrow, curved, and gradually terminating in a very distinct point; their palpi stout, the basal joint very short, the terminal securiform. Prothorax depressed, slightly foliaceous and rounded at the sides, the apex strongly emarginate and much narrower than the base, the latter broadly lobed. Elytra obovate, reticulate, the epipleural fold obsolete. Legs moderately long; femora nearly linear, compressed; posterior tarsi compressed, the basal joint nearly as long as the rest together, the penultimate of all entire. Mesosternum deeply notched for the reception of the prosternal process. Metasternum and abdomen as in *Adelium*, the former, however, rather longer.

This is a very distinct genus, for which at present it is difficult to assign any very near ally, although its habit is that of *Atryphodes*.

Byallius reticulatus. Pl. X. fig. 6.

B. niger, infra et pedibusque nitidis.

Hab. Mountains of Gippsland.

Black; head and prothorax very slightly nitid, minutely punctured, the lateral borders of the latter recurved; scutellum very transverse and glossy; elytra gradually broader from the base, shortly rounded towards the apex, wrinkled with small irregular vermiculate depressions, giving the whole surface a reticulate appearance, the epipleuræ minutely punctured; sterna, abdomen, and legs black, shining; tarsi ferruginous brown, clothed beneath as well as the edge of the lip with rich golden hairs; antennæ with a greyish pubescence towards the tips. Length 9 lines.

Seirotrana proxima.*

S. nigra, convexa, subnitida; prothorace marginibus erosis; elytris fusco-æneis, lineis interruptis elevatis, interstitiis biseriatim punctatis.

Hab. Victoria.

Resembles *S. catenulata*, Boisd., but more convex, entirely subnitid above; the elytra dark brown bronze, with double rows of small simple punctures between the raised interrupted

* Pascoe, Journ. of Entom. ii. p. 483.

lines or tubercles. In *S. catenulata* the middle of the prothorax and elytra is decidedly flattish, the latter a pure dense
black, and between the glossy lines of tubercles opaque ; the
punctures, also in double rows, have each a glossy granule at
the anterior edge. The prothorax in both species is marked
with minute short longitudinal lines, between which the punctures are placed, and the lateral margins are jagged or erose
at their edges. Dr. Howitt says that this new species is the
Victorian representative of *S. catenulata*, whose habitat appears to be confined to the Sydney district. My specimens of
S. proxima are about 6 lines long ; the older species is larger.

Seirotrana crenicollis (Howitt's MS.). Pl. X. fig. 4.

S. planata, brunnescens, subopaca, granulis nitidis instructa, marginibus prothoracis crenatis ; elytris lineis interruptis elevatis, et
granulis minutis seriatim interpositis.

Hab. "Mountains of Victoria."

Light reddish brown, subopaque above, with numerous glossy
granulations of various sizes ; antennæ dark brown ; head
finely granulate ; prothorax longer than broad, nearly flat,
closely covered with small irregular granulations, the margins
pale yellowish brown and crenate ; scutellum nearly hidden
by the overlapping base of the prothorax ; elytra nearly flat,
except towards the apex, where they bend down rather suddenly, a little wider than the prothorax at the base, the sides
subparallel ; the disk with granulations mostly of two sizes,
the largest (of a dark amber-colour) forming interrupted lines,
of which there are four on each elytron ; between these lines
are rows, generally three in number, of small round ones ;
body beneath thickly granulated ; legs light reddish brown,
femora with a broad yellow ring near the apex ; tarsi slender,
filiform. Length 5–6 lines.

A remarkable species, somewhat departing from the normal
form in the longer prothorax and very slender tarsi. *Seirotrana* is distinguished from *Adelium* by its prothorax closely
applied to the elytra, and the shortness of the third antennary
joint, and from *Coripera* by the complete or nearly complete
absence of the epipleural fold ; it is barely to be noticed in the
above species, being indicated by a very narrow line nearly in
the middle of the epipleura.

*Coripera** ocellata* (Howitt's MS.). Pl. X. fig. 5.

C. cupreo-fusca, nitida ; elytris biseriatim impressis, interstitiis an-
nulis oblongis impressis, marginibus disci flavis.

Hab. Mount Macedon (Victoria).

* Pascoe, Journ. of Entom. ii. p. 483.

Dark copper-brown; head finely and irregularly punctured; prothorax with minute shallow punctures, its lateral margins paler; scutellum small, transverse; elytra nearly flat above, each with seven rows of small punctures, the two outer on the epipleural line, the inner bordering the suture, the four intermediate lines placed in pairs, each pair and the sutural and marginal rows separated by a line of oblong impressed rings; the disk bordered with yellowish; body beneath and legs very glossy brown; antennæ and tarsi ferruginous, the latter very slender, filiform. Length 4–5 lines.

Closely agreeing in form with *C. deplanata*, Boisd., but very distinct on account of the peculiar sculpture of the elytra. In my description of the genus *Coripera* the term epipleura was by some oversight used to express the epipleural fold, which, although narrow, is well marked and extends along the whole length of the epipleura; the latter is nearly vertical.

After the following additions have been made to the genus *Adelium* of Kirby*, there remain a few species, the types apparently of as many genera related to it, but differentiated by characters which will not allow them to be conjoined. We find that there are three characters which seem to belong without exception to the *Adelia*, viz. the tarsi tomentose beneath, their penultimate joints subbilobed, and the eyes transverse, narrow, and more or less impinged on by the antennary ridges; a secondary character, because there are cases in which it becomes scarcely recognizable, is the emarginate apex of the prothorax. The subbilobed form of the tarsi is the most permanent of all, and is absent from none of the new genera here recorded. The mentum and lower lip seem subject to considerable modifications; but, after the examination of those of several species, I think it would be unsafe to depend on them alone for generic characters. The subjoined tabular arrangement will give an idea of the diagnoses of the genera :—

Tarsi tomentose beneath.
  Eyes narrow, transverse.
    Anterior tarsi with the three intermediate joints transverse. *Adelium.*
    Anterior tarsi with the three intermediate joints narrow
      and obconic in the female .... *Apasis.*
  Eyes nearly round.. *Brycopia.*
Tarsi pilose beneath.
  Prothorax emarginate at the apex .... *Dystalica.*
  Prothorax not emarginate at the apex.
    Eyes round... *Dinoria.*
    Eyes transverse, narrow .... *Licinoma.*

---

* Trans. Linn. Soc. xii. p. 420.

Adelium plicigerum.

A. nigrum, parum nitidum ; prothorace late transverso, marginibus foliaceis, disco longitudinaliter plicato ; elytris fusco-cupreis, breviter obovatis, interrupte striatis.

Hab. Queensland.

Black, slightly nitid; head irregularly punctured; two transverse wrinkled impressions above the clypeus; prothorax short, the sides strongly rounded, the foliaceous margins very distinct, the disk marked with fine longitudinal, irregular, raised lines; scutellum broadly triangular; elytra of a clear brownish copper-colour, shortly obovate, sharply striate, the striæ interrupted, the alternate intervals between them slightly raised, epipleuræ with scattered punctures; body beneath and legs dark pitchy, impunctate; prosternum and corresponding portion of propectus elevated; antennæ black, the outer joints obconical, the last ovate.   Length 8 lines.

A very distinct species, having the outline of *A. auratum*, but at once distinguished from all other *Adelia* by the sculpture of the prothorax.

Adelium ærarium.

A. viridi-æneum, subnitidum; prothorace transverso, marginibus haud foliaceis, disco creberrime punctato; elytris interrupte costatis.

Hab. Darling Downs.

Greenish bronze, rather nitid; head and prothorax closely punctured, the punctures varying in size and shape, and frequently confluent, the latter transverse, well rounded at the sides and without foliaceous margins; scutellum small, triangular; elytra rather short, the sides but slightly rounded, irregularly costate, the costæ more or less interrupted, the intervals irregular but scarcely punctured, epipleuræ strongly punctured; body beneath dark glossy green, nearly glabrous, but the last abdominal segment punctured; legs and antennæ dark green, clothed with short black sparse hairs, the latter with the outer joints obconic.   Length 7 lines.

Allied to *A. augurale*, but more glossy, the elytra more regularly striate, without granulations, &c.

Adelium pilosum.

A. fusco-cupreum, subnitidum, pilosum; prothorace creberrime punctato, lateribus angulato-rotundatis, haud foliaceis; elytris subcostatis, irregulariter punctato-impressis.

Hab. Lachlan River.

Brownish copper, slightly nitid, everywhere clothed with

short scattered erect hairs, especially on the back; head uneven between the eyes, finely punctured; prothorax transverse,
closely and here and there contiguously punctured, the sides
forming a rounded angle at the middle, not foliaceous, the
apex only slightly emarginate; scutellum indistinct, unicolorous; elytra oblong obovate, subcostate, the intervals with
two rows of irregular punctures, one of the rows with much
larger and more oblong punctures than the other; epipleuræ of
the elytra, and body beneath, glossy purplish black, finely
punctate, or nearly impunctate; legs black, the femora glossy,
with a greenish tinge; antennæ brown, the outer joints obconic,
the last oval. Length 7 lines.

Adelium scutellare.

A. fusco-cupreum, subnitidum, pilosum: prothorace interrupte
punctato; scutello nigro; elytris punctatis et punctato-impressis,
lineisque subelevatis.

Hab. Darling Downs; Brisbane.

Brownish copper, slightly nitid, clothed with short scattered
erect hairs above; head with a few small punctures, uneven
between the eyes; prothorax as in the last, but the punctures
fewer, scattered, and leaving here and there glabrous patches;
scutellum greenish black, broadly triangular; elytra oblong,
rounded at the sides, seriate-punctate, many of the punctures
(two or three together) in oblong impressions, the intervals between the alternate rows slightly raised; epipleuræ of the elytra,
and body beneath, glossy greenish black, the former finely
punctured; legs greenish black, shining, slightly pilose; antennæ brown, with the outer joints elongate obconic, the last
obovate. Length 7–8 lines.

These two species belong to the category of *A. angulicolle*,
Lap., with which my *A. succisum* is probably identical.

Adelium reductum.

A. fusco-cupreum, nitidum; prothorace subtilissime punctato, haud
foliaceo; elytris modice obovatis, leviter seriatim punctatis, punctis
inæqualibus, interstitiis impunctatis; antennis linearibus.

Hab. Brisbane.

Brownish copper, shining; head sparingly and rather finely
punctured; prothorax transverse, the sides well rounded, not
suddenly contracted near the posterior angle, disk very minutely
punctured; scutellum small, rounded behind; elytra not broadly
obovate, seriate-punctate, the punctures small, unequal in size,
some oblong or more deeply impressed, the intervals between
the rows rather wide and impunctate, epipleuræ impunctate:

body beneath and legs dark copper, smooth; tarsi with bright golden-brown hairs; antennæ linear, the joints elongate-ob-conic, pitchy black, ferruginous towards the apex. Length 5½ lines.

Adelium geniale.

A. fusco-cupreum, nitidum; prothorace subtiliter punctato, lateribus subfoliaceis; elytris late obovatis, striato-punctatis, interstitiis subtiliter punctatis; antennis linearibus.

Hab. Clarence River.

Brownish-copper, shining; head and prothorax black, finely punctured, the latter short and transverse, well rounded, broadly margined at the sides, but the margin only slightly foliaceous; scutellum transversely triangular, black; elytra broadly obo-vate, striate-punctate, the striæ well marked, not widely apart, the punctures small and very nearly contiguous, the intervals between the rows slightly convex and rather finely punctured, epipleuræ finely punctured; body beneath and legs pitchy black; tarsi ferruginous beneath; antennæ linear, the joints elongate-obconic, pitchy, ferruginous towards the apex. Length 6½ lines.

This species, as well as the former, belongs to the category of *A. calosomoides,* Kirby. From this the first is distinguished by its narrower form, scarcely punctate prothorax, and the larger and unequal punctures of the elytra; the second, with the same broad outline, has the elytra striated. The next species departs from the *calosomoides*-type in having the antennæ gradually thicker outward, and with shorter joints. The four species have a short curved impressed line on each side of the prothorax.

Adelium neophyta.

A. fusco-cupreum, nitidum; prothorace subtiliter punctato, haud fo-liaceo; elytris subanguste obovatis, striato-punctatis, interstitiis subtilissime punctulatis; antennis apice crassioribus, articulis paulo breviusculis.

Hab. Adelaide; Essendon Plains, Victoria.

Brownish copper (much darker in the Victorian example), shining; head and prothorax black, finely punctured, the latter transverse, moderately well rounded at the sides, not foliaceous; scutellum black, broadly triangular; elytra rather narrowly obovate, striate-punctate, the striæ broad and shal-low, the punctures rather small and not nearly contiguous, the intervals of the striæ slightly convex and very minutely punc-tured, epipleuræ glabrous, nearly impunctate; body beneath and legs smooth, pitchy black, tibiæ and tarsi with ferruginous hairs; antennæ a little thicker outwardly, the joints obconic, not elongated, the third equal to the fourth and fifth together, pitchy, with scattered short hairs. Length 4½ lines.

A. brevicorne, Blessig, judging from the figure he has given*, appears to be a much broader species, with the prothorax much less narrowed at the apex; in the description the latter is said to be twice as broad as long.

Adelium ancilla.

A. cupreum, sat nitidum; prothorace subtiliter punctato, angulis posticis productis; elytris irregulariter seriatim impresso-punctatis.

Hab. Darling Downs.

Copper-brown, rather nitid; head sparingly punctured; clypeus rounded at the apex, its suture somewhat indistinct, but the groove at the base of the antennary ridges well marked; prothorax transverse, much narrower than the elytra, finely and rather remotely punctured, broad at the base, the apex narrowed, sides strongly rounded, posterior angles produced directly outwards; scutellum transversely triangular, its apex rounded; elytra broadly obovate, convex, seriate-punctate, the punctures irregularly impressed, oblong or round, here and there one, two, or three together, the intervals of the rows broad, impunctate, and more or less uneven from the impressed sides of the punctures; epipleuræ of the elytra, body beneath, and legs glossy reddish copper, sparingly and finely punctured; antennæ more than half the length of the body, slightly thicker outwards, glossy copper at the base, gradually becoming ferruginous and opaque, third joint nearly as long as the next two together, the last joint a little larger than the preceding one, and somewhat semicircular. Length 5½ lines.

Differs from *A. cisteloides*, Er. (?*A. helopoides*, Boisd.), *inter alia*, in its longer antennæ, and in the greater breadth of the base, and the produced posterior angles of the prothorax.

Adelium repandum.

A. cupreum, subnitidum; prothorace creberrime punctato, punctis magnis rarissime dispersis; elytris seriatim punctatis, interstitiis alternis interrupte subcostatis.

Hab. Brisbane.

Copper-brown, a little nitid; head rather finely punctured; clypeus truncate at the apex, its suture going off near the antennary ridges, no branch groove; prothorax moderately transverse, not so broad as the elytra, sides well rounded, posterior

* Horæ Soc. Ent. Rossicæ, fasc. i. p. 101, taf. 3, fig. 2.

angles not produced, base emarginate, rather close to and slightly overlapping the elytra, closely and minutely punctured, a few large punctures irregularly dispersed among them; scutellum very short and transverse; elytra obovate, irregularly seriate-punctate, punctures small, not crowded, the intervals between the rows broad, the alternate ones with slightly raised interrupted lines, epipleuræ with a few minute scattered punctures; body beneath and legs dark greenish brown, very glossy, the middle of the abdominal segments finely corrugated; antennæ rather short, copper-brown, thicker outwards, third joint a little longer only than the fourth, the last ovate, much longer than the tenth. Length 5½ lines.

A distinct species; in the closeness of its prothorax to the elytra, and also in habit, slightly approaching the genus *Coripera*.

Adelium scytalicum.

A. fusco-cupreum, pernitidum; prothorace nigro, lævissimo; elytris seriatim punctatis, punctis inæqualibus.

Hab. Swan River.

Brownish copper, very nitid; head and prothorax black, the former minutely punctured, the latter very smooth and glossy, rather transverse, the sides well rounded and not foliaceous, the base and apex of equal breadth; scutellum nearly semicircular; elytra oblong, slightly rounded at the sides, seriate-punctate, the punctures unequal in size, the intervals but very slightly convex; epipleuræ of the elytra, legs, and body beneath very smooth and glossy; tarsi ferruginous; antennæ dark brown, eighth, ninth, and tenth joints triangular, dilated on one side, the eleventh obliquely ovate, larger than the preceding one. Length 5 lines.

I have but one example of this very distinct species; it is probable that the peculiarity of the antennæ is sexual.

Adelium orphana.

A. cupreum, nitidum; prothorace subtiliter punctato; elytris striato-punctatis.

Hab. Yankee Jim's Creek, Victoria.

Glossy copper-brown; head finely punctured; clypeus very slightly emarginate at the apex, its suture moderately arched; prothorax transverse, nearly as broad as the elytra, a little rounded at the sides, minutely punctured, posterior angles not produced; scutellum transversely triangular; elytra subparallel at the sides, punctate-striate, punctures rather small and approximate, intervals of the striæ thickly punctured, epipleuræ finely punctured; body beneath and legs glossy copper; tarsi fulvous; antennæ ferruginous, gradually thicker outwards; last joint larger than the tenth, somewhat semi-circular. Length 4½ lines.

Very like an *Amara* in habit; narrower, more parallel at the sides, and more glossy than any of the others.

The three following species have a more slender form than the *Adelia* generally: the prothorax is also less transverse and only slightly emarginate at the apex, and the eyes are broader and less impinged on by the antennary ridges. The third species has the prothorax nearly as broad at the base as at the apex, while in the first two it is very much narrower. They lead to a certain extent to *Apasis*, from which, however, they are separated by the characters of their anterior tarsi.

Adelium steropoides.

A. gracile, æneum; prothorace apice parum emarginato, basi angustiore; elytris punctato-striatis.

Hab. Victoria.

Brassy, nitid; head concave and thickly punctured between the antennary ridges, the front with a slightly bilobed gibbosity; clypeus deeply emarginate; prothorax rather broader than long, the sides well rounded, narrowed at the base, very minutely punctured; scutellum triangular; elytra oblong, slightly rounded at the sides, moderately convex, punctate-striate, the punctures nearly contiguous, the intervals of the striæ narrow, convex, and impunctate; epipleuræ of the elytra, body beneath, and legs glossy copper-brown, with minute scattered punctures; tarsi and outer joints of the antennæ ferruginous. Length 6½ lines.

Adelium ruptum.

A. gracile, piceo-fuscum; prothorace apice parum emarginato, basi angustiore; elytris æneis, striatis, striis interruptis.

Hab. Yankee Jim's Creek.

Pitchy brown, nitid; head concave between the antennary ridges, rather thickly punctured, front slightly raised between the eyes; clypeus tinged with steel-blue, deeply emarginate, the upper lip very short and narrow; prothorax slightly transverse, well rounded at the sides, narrowed at the base, very minutely punctured; scutellum rather narrowly triangular; elytra oblong, slightly rounded at the sides, a little depressed, striate, the striæ more or less interrupted, the intervals of the striæ flattish and nearly impunctate, epipleuræ indistinctly punctured; body beneath and legs dark brown, glossy; tarsi and outer joints of antennæ ferruginous. Length 7 lines.

Adelium commodum.

A. gracile, nigrum; prothorace apice parum emarginato, basi haud angustato; elytris æneis, tenuiter subpunctato-striatis.

Hab. Tasmania.

Black, subnitid; head scarcely punctured, flattish in front and above the eyes; clypeus strongly emarginate, somewhat ferruginous, as well as the upper lip; prothorax as long as broad, apex slightly emarginate, sides moderately rounded, base rather broad, but less so than the apex, the disk very slightly convex and scarcely punctured; scutellum transverse; elytra slightly rounded at the sides, finely striate, the striæ with traces of punctures only, the intervals narrow, with an indistinct punctuation; epipleuræ of the elytra, body beneath, and femora glossy reddish brown, with minute shallow punctures; tibiæ reddish ferruginous; tarsi and antennæ paler, inclining to fulvous.   Length 5 lines.

APASIS.

Mentum angulis anticis rotundatum.
Prothorax apice truncatus.
Tarsi ant. in fœm. art. tribus intermediis obconicis; omnes subtus tomentosi.

The type of this genus has a very different appearance from any of the species of *Adelium*; and therefore, in the absence of any very salient differential character, I have been led to attach some importance to the peculiar form of the intermediate joints of the anterior tarsi of the female; for in the male they are transverse, as in both sexes of *Adelium**, but more dilated.

I owe my specimens, as well as all the new allied forms here described, to my kind friend Dr. Howitt, to whom I dedicate the species.

Apasis Howittii.   Pl. XI. fig. 7, ♂.

A. atra, nitida; tarsis palpisque fulvis; elytris striatis.

Hab. Victoria.

Black, shining; head nearly impunctate, very hollow between the antennary ridges in the line of the clypeal suture, a transverse groove in front above the eyes; clypeus strongly emarginate, upper lip large and prominent; prothorax very glabrous, finely and sparsely punctured, about equal in length and breadth, convex, rounded at the sides, the margins with a

---

* The anterior and frequently the intermediate tarsi are more dilated in the males of *Adelium* than in the females.

narrow raised border; scutellum transverse; elytra oblong
oval, a little broader than the prothorax, slightly rounded at
the sides, striate, the striæ and the spaces between them impunctate, scutellar stria nearly obsolete; epipleuræ of the
elytra, body beneath, and legs pitchy brown, very smooth and
glossy; tarsi and palpi fulvous; antennæ a little thicker outwards. Length 10 lines.

LICINOMA.

Mentum angulis anticis rotundatum.
Tarsi subtus leviter pilosi.
Prothorax apice haud emarginatus.

In other respects, except that the eye is broader, this genus
resembles *Adelium*, with the habit of some of the smaller
Feroniæ.

Licinoma nitida.

L. cuprea, nitida; elytris punctato-striatis; tarsis fulvis; antennis
ferrugineis.

Hab. Mount Macedon, Victoria.

Copper-brown, shining, finely punctured; head convex between the antennary ridges, sparsely punctured; clypeus
emarginate at the apex; prothorax nearly as long as broad,
the sides slightly rounded, a little narrowed at the base; scutellum small and indistinct; elytra oblong, very moderately
rounded at the sides, scarcely broader than the prothorax, delicately punctate-striate, the intervals of the striæ flattish,
sparingly punctured; epipleuræ of the elytra, body beneath,
femora, and tibiæ glossy reddish brown, sparsely punctured;
tarsi fulvous; antennæ ferruginous, thicker outwards, the last
joint large and as long as the two preceding together. Length
4 lines.

BRYCOPIA.

Oculi prominuli, subrotundati.
Mentum angulis anticis rotundatum.
Prothorax apice haud emarginatus.

The principal differentiating character in this genus is the
prominent and nearly circular eye. The simple clypeal suture
may probably also be taken as a generic character. The tarsi
are closely tomentose beneath, as in *Adelium*.

Brycopia pilosella.

B. breviter et sparse pilosa; capite prothoraceque violaceo-nigris;
elytris cupreis, punctato-striatis.

Hab. Mount Macedon, Victoria.

Shining above, with short erect scattered hairs; head and prothorax violet-black, coarsely punctured, the clypeal suture not sending a branch along the base of the antennary ridge; sides of the prothorax well rounded anteriorly, then contracting more gradually to the base; scutellum triangular; elytra oblong oval, coarsely punctate-striate, the intervals between the striæ impunctate, epipleuræ scarcely punctured; body beneath reddish pitchy, punctured; legs and antennæ pale ferruginous, the last joint of the latter rounded, a little larger than the preceding one.   Length 3 lines.

DINORIA.

Oculi parvi, rotundati.
Tarsi subtus pilosi.
Prothorax apice haud emarginatus.

Very similar to *Brycopia*, and only to be distinguished by the pilose tarsi. The clypeal suture is also simple.

Dinoria picta.

D. cuprea, nitida; elytris marginibus fulvis.

Hab. Tasmania.

Copper-brown, shining; head coarsely punctured, the clypeus forming a prominent fold above; prothorax transverse, roughly and not closely punctured, rounded at the sides, more narrowed behind the middle, the posterior angles prominent, but not acute; scutellum narrowly triangular; elytra obovate, closely and strongly punctate-striate, the intermediate intervals more elevated, the margins of the disk and apex fulvous; body beneath and femora at the base dark glossy brown, sparsely punctured; rest of the femora, tibiæ at the apex, and tarsi clear fulvous; palpi and antennæ yellowish ferruginous, the latter a little thicker outwards, the last joint oval, nearly equal to the two preceding together.   Length 3 lines.

DYSTALICA.

Oculi angustati.
Tarsi subtus pilosi.
Prothorax apice emarginatus, lateribus crenatus.

In habit resembling *Adelium porcatum* more than anything else in the subfamily.

Dystalica homogenea.

D. subparallela; capite prothoraceque nigris; elytris æneis, fortiter punctato-striatis.

Hab. Swan River.

Head and prothorax closely and rather coarsely punctured; clypeal suture strongly arched, sending back on each side a shallow groove terminating near the upper edge of the eye; prothorax much broader than long, convex, the sides rounded and remotely crenate; scutellum narrowly triangular; elytra oblong, the sides nearly parallel, about the width of the prothorax, strongly punctate-striate, the punctures approximate, intervals between the striæ narrow and very convex, epipleuræ coarsely and rather closely punctured; body beneath and legs greenish black, glossy, slightly punctured; antennæ with the third joint elongate, fourth to tenth equal and obconic, the last oval, not larger nor longer than the tenth. Length 8 lines.

Omolipus lævis.

O. ater, nitidus; antennis tarsisque ferrugineis; elytris subtiliter seriatim punctatis.

Hab. Cape York.

Black, shining; head and prothorax very minutely punctured, the latter transverse, well rounded at the sides, the base broader than the apex; scutellum very small, triangular; elytra shortly ovate, seriate-punctate, the punctures very small and invisible to the naked eye; body beneath and legs very glossy; the antennæ, palpi, and tarsi ferruginous; claw-joint very stout. Length 6 lines.

Omolipus gnesioides.

O. ater, nitidus; prothorace antice gibbosulo; elytris fortiter seriatim punctatis, punctis oblongis.

Hab. Port Denison.

Black, shining; head very minutely punctured, punctures somewhat scattered, much more crowded on the clypeus; prothorax also minutely punctured, somewhat compressed, and becoming slightly gibbous anteriorly, the sides moderately rounded; scutellum small, transverse, rounded behind; elytra rather narrowly ovate, seriate-punctate, the punctures oblong and strongly impressed; body beneath and legs very glossy; antennæ and tarsi black. Length 4 lines.

Omolipus (Pascoe, Journ. of Entom. i. p. 127) is allied to the European genus *Misolampus*, from which it may be at once distinguished by the presence of a scutellum and the hooked inner maxillary lobe. The species are all of an intense black colour, more or less glossy; and, in addition to the characters given of the genus, it may be stated that the claw-joint is unusually stout, and the epipleura gradually narrows posteriorly and disappears a little way from the apex.   The other two species may be diagnosed as follows :—

Omolipus corvus, Pasc. *l. c.*—Ater, nitidus ; prothoracis basi apice angustiore ; elytris fortiter seriatim punctatis.

Hab. Brisbane *.

Omolipus socius, Pasc. (Ann. & Mag. Nat. Hist. ser. 3. ix. p. 463).— Ater, nitidissimus; prothoracis basi apice latiore ; elytris fortiter seriatim punctatis, punctis distantibus.

Hab. Lizard Island.

ECTYCHE.

Subfamily *AMPHIDORINÆ*.

Clypeus a fronte discretus, antice paulo rotundatus.
Tibiæ anticæ apice dilatatæ, oblique truncatæ.
Processus intercoxalis angustatus, apice rotundatus.

Head rather short, inserted into the prothorax as far as the eyes, regularly convex in front; the clypeus large, a little rounded anteriorly, separated from the front by a strongly arched suture. Eyes narrow, entire. Antennæ slightly thicker outwards, the third joint longer than the others, the fifth to the tenth more or less ovate, submoniliform, the last larger and oblong. Mentum pedunculate, trapezoidal, the anterior border slightly biemarginate ; labium small, membranous, transverse. Maxillary lobes narrow, the inner hooked. Maxillary palpi large, strongly securiform ; the labial short, thick, approximate at the base. Prothorax transverse, convex, apex rather slightly emarginate, sides rounded but broadly emarginate at the posterior angle, the emargination with a tooth in the middle. Elytra ovate, convex, the shoulders rounded ; epipleura broad at the base, gradually narrower and almost obsolete at the apex. Femora strongly clavate ; anterior tibiæ toothed along the outer margin, gradually thicker below, the apex obliquely truncate and terminating in a stout spur inwardly, the intermediate and posterior linear, the edges round their cotyloid cavities spinous ; tarsi slender, setose beneath ; the basal joint of the posterior moderately elongate. Prosternum abruptly elevated, rounded anteriorly and posteriorly. Mesosternum abrupt and a little excavated in front. Metasternum very short. Abdomen with the third and fourth segments membranous at their edges. Body, with the legs and antennæ, covered with long flying hairs.

After Dr. Leconte, I have taken *Amphidora* as the type of a subfamily perfectly distinct from the Adeliinæ, in which

---

* Not Melbourne, as erroneously stated in the 'Journal of Entomology.'

M. Lacordaire places it, although with some doubt. The sub-family forms to a certain extent an exception to the cognate groups in regard to the tarsi, the pubescence beneath being "very coarse, sometimes almost spinous;" in *Ectyche* it is completely setose (or spinous). The Amphidorinæ hitherto have been exclusively Californian and Chilian; and, notwithstanding there are so many points of agreement between the latter and the Australian beetle-faunas, it was not until after a long examination that I ventured to consider this genus one of its members. All the essential characters, however, are the same, the intercoxal process, very broad in *Amphidora* itself, is considerably less so in *Stenotrichus*; and we have seen that the vestiture of the tarsi is variable.

I owe my specimen to the Rev. George Bostock, of Free-mantle.

Ectyche erebea. Pl. XI. fig. 1.

E. oblonga, nigra, opaca; elytris striato-punctatis, interstitiis crebre punctatis.

Hab. Freemantle.

Black, opaque, everywhere above covered with long, erect, slender, black hairs; head, upper lip, and prothorax closely and finely punctured; scutellum minute, punctiform; elytra about three times the length of the prothorax, striate-punctate, the intervals closely and rugosely punctured; breast glabrous, closely punctured; abdomen coarsely punctured, hairy. Length 2 lines.

The following species is closely allied to *Ectyche*, but differs in the character of the tibiæ, which are all of the same form and toothed (or rather, perhaps, shortly spined) externally. It is a much smaller species; and my specimen, which I owe to Mr. Odewahn, of Gawler, having been carded, the gum(?) used has such a tenacious property that it is impossible to get rid of it so as to be able to examine the different organs satisfactorily. I record it here principally to call the attention of Australian entomologists to the subject. The occurrence of two such closely allied species so far apart suggests the probability that these are by no means such isolated forms as they now appear to be. It is not unlikely that they are ants'-nest insects.

Ectyche? nana.

E.? breviter ovata, nigra, opaca; elytris subnitidis, crebre punctatis, interstitiis rugosis.

Hab. Gawler.

Shortly ovate, opaque black, but the elytra slightly glossy, covered above with long black erect hairs; head and prothorax finely and closely punctured; clypeus not distinct from the front; prothorax transverse, convex, rounded at the sides and anterior angles, the posterior acuminate; scutellum inconspicuous; elytra scarcely broader than the prothorax, subnitid, the punctures mostly irregular, or with slight indications of rows, crowded, the intervals rugose; body beneath dark brown, closely punctured; antennæ and legs ferruginous; tibiæ slightly compressed, gradually dilated downwards, the outer edge shortly spined; tarsi with longish hairs beneath. Length 1¾ line.

BRISES.

Subfamily *Cœlometopinæ.*

Caput ad oculos retractum.
Maxillæ lobo interiore mutico.
Prothorax lateribus foliaceis.

Head transverse, inserted into the prothorax as far as the eyes; antennary ridges dilated; clypeus broad, separated from the front by a slightly arched line, strongly emarginate in front. Eyes transverse, entire. Antennæ small, thicker outwards; third joint elongate; fourth, fifth, and sixth obconic; seventh to the tenth submoniliform, the last obovate. Mentum, as well as the labium, transverse, broader and truncate anteriorly. Maxillary lobes small, the inner short and unarmed. Palpi gradually thicker outwards; the maxillary with a short basal joint, second as long as the two following together, the last narrowly triangular; the labial with a basilateral insertion. Prothorax transverse, the apex strongly emarginate, sides foliaceous and recurved, disk scarcely convex, the base subtruncate, with the posterior angles narrowly produced. Elytra shortly ovate, broader than the prothorax at their base, shoulders round; epipleuræ gradually narrowing posteriorly. Legs rather feeble; femora slightly thickened, the anterior with trochanters; tibiæ linear, shortly spurred; tarsi clothed beneath with long, stiff hairs, the middle and posterior with the claw-joint as long, or nearly as long, as the preceding joints together. Prosternum elevated, produced above. Mesosternum V-shaped. Metasternum short. Intercoxal process narrowly triangular, obtuse anteriorly. Abdomen with the fourth segment very short, and with the third incurved at the sides.

This is another of those special forms in which Australia is so prolific; and therefore there is little to be said respecting its affinities. As may be supposed, it differs in some respects from the characters of the Cœlometopinæ as laid down by M. Lacordaire. Many species of this subfamily are Californian, where, according to Dr. Leconte, they are found under the bark of trees. We are ignorant of the habits of the Australian species.

Brises trachynotoides. Pl. XI. fig. 5.

B. nigro-fusca, opaca; elytris granulatis, punctatis, singulis bicostatis.

Hab. Champion Bay.

Opaque blackish brown; head and prothorax finely and very closely granulate, the granulations more or less confluent; scutellum transverse, pointed at the tip; elytra moderately convex, irregularly punctured, with the intervals granulate, each elytron with two very marked costæ not reaching to the apex; body beneath pitchy brown, finely but obscurely punctured; antennæ and legs dark ferruginous, covered with scattered stiffish hairs. Length 7½ lines.

ASPHALUS.

Subfamily CŒLOMETOPINÆ.

Caput ad oculos retractum.
Maxillæ lobo interiore hamato.
Tarsi omnes art. ultimo cæteris simul sumptis longiore.

Head transverse, inserted into the prothorax as far as the eyes; clypeus separated from the front by an arched line, slightly emarginate anteriorly; labrum broadly transverse, porrect. Eyes transverse, nearly entire. Antennæ rather short, gradually thicker outwards, the third joint a little longer than the second and fourth, and all, as far as the seventh, obconic; eighth, ninth, and tenth broader and shorter, the last larger than the preceding, round and a little depressed. Mentum shortly pedunculate, hexagonal, winged*; labium very transverse, subtrilobed. Maxillary lobes—inner narrow, gradually terminating in a strong hook; outer short, broad, somewhat triangular. Maxillary palpi stout, the last joint securiform; labial short, last joint large, cup-shaped. Prothorax convex, broader than long, sides rounded, terminating posteriorly in a strongly produced acute angle; apex deeply and broadly emarginate, base bisinuate. Elytra ovate, as broad as the prothorax, convex; epipleuræ entire, gradually narrowing from the shoulder to the apex. Legs stout, the posterior

* Mr. F. Bates (Trans. Ent. Soc. 1868, p. 259) proposes by this word to designate that "peculiar form of mentum composed of a central portion large and convex and two smaller flat pieces (wings) situated on each side at the back." These wings appear to be the "lateral lobes" of Dr. Leconte. The presence of these lobes differentiates *Nyctobates* from *Iphthimus.*

longest; femora gradually thickened, furnished with trochanters; tibiæ shortly spurred, intermediate and anterior
arched; tarsi short, entire, the claw-joint longer than the rest
together. Prosternum broad, produced behind. Mesosternum
broadly V-shaped. Metasternum very short. Intercoxal process small, quadrate. Abdomen with the third and fourth
segments strongly incurved at the sides.

In habit resembling *Pedinus*, with which I at first thought
this genus might possibly be connected; but its true place is
with the Cœlometopinæ. Mr. F. Bates has already placed his
two Australian genera *Hypaulax* and *Chileone*, dismembered
from *Nyctobates*, in this subfamily; but these are very different
in appearance from *Asphalus*. There is a considerable depression on the throat of the species here described, which represents the grooves of *Hypaulax* and *Cælometopus*. The lower
lip is also remarkable, inasmuch as the central lobe appears to
be corneous, whilst the lateral ones are membranous.

Asphalus ebeninus.   Pl. XI. fig. 3.

A. aterrimus, nitidus, lævis; elytris leviter punctato-striatis.

Hab. Clarence River.

Deep black, smooth and shining; antennæ and tarsi ferruginous; head and prothorax very minutely punctured, the
latter with the sides rather more broadly margined anteriorly
than posteriorly; scutellum very short, transverse; elytra very
convex, faintly punctate-striate, the epipleura at its junction
with the disk forming a prominent line, especially anteriorly;
body beneath more or less finely corrugated. Length 8 lines*.

PROMETHIS.

Subfamily *TENEBRIONINÆ*.

Caput exsertum, pone oculos collo cylindrico contractum.
Prothorax angulis anticis productis, rotundatis; marginibus integris.
Tibiæ haud calcaratæ; *tarsi* postici validi, breviusculi.

The type of this genus is "*Upis (Iphthimus) angulatus*," Er.†,

---

* Mr. F. Bates, as we have already noticed, having withdrawn several
species previously placed with *Nyctobates*, to form his two genera *Hypaulax*
and *Chileone*, which he places in Cœlometopinæ, it will be necessary to
constitute another for my *N. feronioides*. This genus, which I propose to
name *Hydissus*, differs essentially from both the above in having the
penultimate joint of all the tarsi subbilobed; it has no grooves behind the
mentum; and the epipleural line terminates at the shoulder, this raised
and strongly marked line, which in *Hypaulax* is continuous with the
basal, being interrupted, the basal line turning backwards and running
down for a short distance inside and parallel to the other.

† Wiegmann's Archiv, 1842, i. p. 174. It is found in Victoria as well
as in Tasmania.

a species remarkable for the bearded mentum of its males—a peculiarity which does not appear to be anything more than specific. This genus is differentiated both from *Upis* and *Iphthimus* by the form of its prothorax, and its entire margins when compared with the latter,—to which, as a secondary character, may be added the sculpture of its elytra. The first of the two species described below has been long known in collections ; and in my own it formerly stood as a *Baryscelis*, an unpublished name of Dr. Boisduval. *Iphthimus niger*, Blessig, appears to be in some respects intermediate between the two following.

Promethis lethalis.

P. nigra, subnitida ; prothorace basi angustiore ; elytris postice latioribus, fortiter punctato-striatis, interstitiis convexis.

Hab. Queensland.

Black, shining ; head minutely punctured ; clypeus slightly emarginate at the apex, separated from the front by a fine transverse line bent downwards at the sides ; prothorax very finely punctured, longer than in *P. angulata*, gradually narrowing towards the base, strongly canaliculate on the disk, with two impressed spots on each side ; scutellum semicircular ; elytra much broader than the prothorax at the base, and gradually widening posteriorly, rounding towards the apex, deeply punctate-striate, the punctures indistinct, the intervals raised and very convex ; beneath glossy black ; first three segments of the abdomen finely and thickly punctured ; legs pitchy ; antennæ ferruginous, scarcely extending to the middle of the prothorax. Length 13 lines.

A much larger species than *P. angulata*, but with shorter antennæ proportionally, more nitid, a longer prothorax contracted behind, and strongly striated elytra, which are considerably broader posteriorly. In the following species the elytra are nearly parallel, and the prothorax has the apex and base of the same breadth.

Promethis quadricollis.

P. nigra, subnitida ; prothorace transversim subquadrato ; elytris subparallelis, punctato-striatis, interstitiis modice convexis.

Hab. Swan River.

Resembles the last, but with head and prothorax much less finely punctured, the latter very much more transverse, not narrower at the base, slightly canaliculate ; elytra nearly parallel at the sides, punctate-striate, the striæ broad and shallow, the punctures large, intervals of the striæ moderately convex ; abdomen very minutely punctured, the second and third segments with a series of short longitudinal ridges at the base. Length 9 lines.

It will be necessary to form a new genus for the reception of *Upis cylindrica*, Germ.[*], which, as M. Lacordaire justly observes, is more related to *Menephilus* than to *Upis*. It is a very distinct form, for which I propose the name of

ŒCTOSIS.

Oculi angustati, infra acuti.
Prothorax angulis posticis rotundatis.
Epipleura postice defecta.

It is a less depressed form than *Menephilus*, and has on each side between the base of the mandible and the eye a prominent fold, as in *Iphthimus*; and it is this apparently which gives the latter its peculiar form. The prosternum is recurved behind, and terminates in a short triangular process. The absence of the epipleura towards the apex is also characteristic of *Dechius*, Pasc.[†], another Australian genus of this subfamily, but which is notwithstanding more allied to *Tenebrio*, as it appears to me, on account of its spurred tibiæ. My specimen is from the Darling River.

MENERISTES.

Subfamily *TENEBRIONINÆ*.

Tibiæ calcaratæ; *femora* incrassata.

This genus differs only in the above characters from *Menephilus*, Muls. The type I have received from Dr. Howitt, under the name of "*Baryscelis laticollis*, Boisd." That genus was never published; but, according to a note of M. Lacordaire's, it belongs to the Cœlometopinæ, and therefore cannot be this. In the British Museum the same species is labelled "*Tenebrio australis*, M'L. (Boisd.)." The descriptions of Dr. Boisduval in the 'Voyage de l'Astrolabe' are very short, varying from five Latin words to five-and-twenty, the latter exceptional; and these are followed by a strictly literal French translation. With the vague ideas of genera common thirty years ago, and even later, the generic name affords scarcely any clue, and it is only by a sort of tradition that we are able

---

[*] Linn. Entom. iii. 198.
[†] Journ. of Entom. ii. p. 455. Mr. F. Bates (Trans. Ent. Soc. 1868, p. 265) contradicts my statement as to the absence of the hook on the internal maxillary lobe of *Dechius aphodioides*. This part has since been examined by Messrs. Smith and C. Waterhouse, of the British Museum, who agree with me that it does not possess a vestige of such a peculiarity.

to accept at all many of Dr. Boisduval's names*. The types, many of them at least, seem to have been lost. I retain the name of "*laticollis*," as it is sufficiently distinctive, and, should it hereafter be found to be the species so designated by Dr. Boisduval, there will be no alteration.

Meneristes laticollis. Pl. XI. fig. 2.

M. niger, nitidus ; sutura clypeali valde impressa ; prothoracis angulis anticis et posticis productis ; tibiæ anticæ valde arcuatæ.

Hab. Victoria.

Black, shining ; head glossy, finely and closely punctured ; clypeal suture arched, strongly impressed ; prothorax minutely punctured, anterior angles produced, subacuminate, posterior terminating in a long acute angular process ; scutellum curvilinearly triangular ; elytra nearly parallel, coarsely punctate-striate, punctures slightly quadrate, very close together, intervals between the striæ very narrow ; body beneath and legs glossy black ; anterior tibiæ equal in length to the intermediate, strongly curved. Length 9 lines.

Meneristes intermedius.

M. niger, nitidus ; sutura clypeali impressa ; prothoracis angulis minus productis ; tibiis anticis (♀) vix arcuatis.

Hab. Gawler.

Black, shining ; head opaque, finely punctured ; clypeal suture arched, moderately impressed ; prothorax as in the last species, but the angles, especially the posterior, less produced ; scutellum triangular ; elytra broader in proportion, less glossy, and less strongly punctured ; body beneath black, shining ; legs glossy reddish pitchy ; anterior tibiæ in the male strongly arched ; shorter, and nearly straight, at least on the outer edge, in the female ; the former only with the tarsi dilated. Length 8 lines.

A stouter insect comparatively than the last, and differing in the form of the anterior tibiæ. In both species there is a deep fovea on each side of the prothorax at the base.

Meneristes servulus.

M. niger, nitidus ; sutura clypeali vix impressa ; prothoracis angulis posticis productis (♂); tibiis anticis in mare longioribus, arcuatis, apice penicillatis.

Hab. "Tasmania to Queensland."

* Under the name of *Mallodon australis*, Boisd., for example, M. Lacordaire says he found "several species, belonging to different genera, in collections." (Gen. viii. p. 111, note.)

Black, shining; head glossy, very minutely punctured; clypeal suture marked by a smooth arched line only; two small impressed curved lines between the eyes; prothorax longer in proportion to the width, very smooth, anterior angles rounded, the posterior narrowly produced in the male; scutellum triangular; elytra narrower anteriorly and not much broader than the prothorax at the base, the greatest width a little distance from the apex in the male, the base broader in the female, punctate-striate as in the last; body beneath and legs pitchy; anterior tibiæ in the males much longer than the rest, strongly arched, and having a tuft of golden hairs at the apex; in the females shorter, less arched, and without the tuft at the apex. Length $6\frac{1}{2}$ lines.

Ephidonius.

Subfamily *Tenebrioninæ*.

Caput exsertum, rhomboideum, pone oculos elongatum.
Tibiæ fortiter calcaratæ; *tarsi* subtus subnudi.

Head exserted, rhomboidal, broad in front, gradually narrowed behind the eyes; clypeus widely emarginate at the apex, its suture nearly straight, except at the sides. Eyes small, rather narrow, transverse. Antennæ slender; third joint longest; fourth, fifth, and sixth shorter, obconic, nearly equal in length; seventh to tenth more or less obovate; the last ovate, pointed, scarcely longer than the tenth. Mentum trapezoidal, broadest and truncate anteriorly. Labium corneous in the middle, with two rounded membranous lobes at the sides; its palpi elongate, a little thicker outwards. Maxillæ with two short lobes, the inner narrow and unarmed; maxillary palpi with the first joint very short, second long, the last obconic, truncate at the apex. Prothorax transverse, slightly emarginate anteriorly, anterior angles not produced, sides rounded, without a raised border, and terminating in well-marked posterior angles, base broad and subtruncate. Elytra ovate, slightly convex, broader than the prothorax. Femora sublinear; tibiæ strongly spurred; tarsi slender, gradually longer from the anterior, nearly naked beneath, except a few setæ at the tips. Prosternum narrow, declivous; mesosternum V-shaped; metasternum moderately elongate. Abdomen with the fourth segment very short, its sutural edge arched.

The general appearance of the insect forming the type of this genus is more nearly that of *Iphthimus italicus* than any other known to me. The vestiture of the tarsi, however, and the presence of spurs to the tibiæ is sufficiently distinctive; the former character, indeed, may lead to the doubt of its belonging to the Tenebrioninæ at all; but in this case I believe it is exceptional.

I am indebted for my specimens to Johannes Odewahn, Esq.

Ephidonius acuticornis.   Pl. XI. fig. 6.

E. niger, capite prothoraceque nitidis; elytris opacis, seriatim et leviter punctulatis.

Hab. Gawler, South Australia.

Black; head and prothorax finely punctured, shining, the former from the clypeus backwards smooth and convex; base of the prothorax close to the elytra, but below their level; scutellum triangular; elytra finely seriate-punctate, the suture thickened into a line, three other lines also on each elytron placed on the intervals of every four rows of punctures; body beneath and legs shining pitchy brown; antennæ reddish brown.   Length 9 lines.

TANYLYPA.

Subfamily *BORINÆ.*

Oculi transversi.
Maxillæ lobo interno inermi.
Tibiæ arcuatæ.

Head exserted, small, gradually narrower behind the eyes; clypeus separated from the front by a short arched suture. Eyes rather narrow, transverse.   Antennæ a little thicker outwards; the basal joints more or less obconic, the eighth to the tenth transverse, the last rounded.   Mentum trapezoidal; labium short, transverse, corneous.   Maxillary lobes short, the inner narrow, unarmed, the outer broadly triangular.   Maxillary palpi stout, broadly dilated upwards; the labial distant at the base, with the last joint very large, cup-shaped, obliquely truncate.   Prothorax longer than broad, narrowed and truncate at the apex, sides rounded, posterior angles acute, the base truncate.   Elytra elongate, parallel, not broader than the prothorax, rounded at the apex; epipleura narrow and nearly equal throughout, but expanding as it ascends to the shoulder. Femora stout; tibiæ strongly curved; tarsi short, the last joint as long as the rest together.   Anterior coxæ transverse.   Prosternum slightly elevated, rounded behind.   Mesosternum short, V-formed.   Metasternum elongate.   Intercoxal space very narrow, short, triangular.   Abdomen with five segments, all nearly equal in length and with corneous edges.

Allied to *Boros*, Herbst, a genus placed with the Pythonidæ by Dr. Leconte*, on account of its anterior cotyloid cavities

* Class. Col. N. Am. p. 255.

being open behind. The same authority also credits them with conical anterior coxæ. I do not know the American species; but in *B. Schneideri* they are slightly transverse*, and they are still more so in the present genus. Although I cannot agree to separate *Boros* from the Tenebrionidæ, as Dr. Leconte and M. C. G. Thomson have done, yet it seems desirable to keep them apart from *Calcarinæ*, with which they do not appear to be very intimately connected.

Tanylypa morio. Pl. XI. fig. 4.

T. nigra, nitida; prothorace basi trifoveolato; elytris seriatim punctatis.

Hab. Tasmania.

Black, shining; head and prothorax very finely punctured, the latter with three very distinct foveæ at the base; scutellum semicircular; elytra rather strongly punctured in closely approximate rows, the sutural row diverging near the scutellum, a very short one taking its place; body beneath and legs dark pitchy, smooth and shining; antennæ glossy ferruginous; fore tibiæ with a delicate fringe of hairs within.   Length 6 lines.

The three following appear to be degraded Tasmanian forms of *Cestrinus*, Er.†, and are closely allied; they are narrower and more feebly constructed, and the prothorax wants the expanded margin. *Opatrum piceitarse*, Hope, belongs to this genus; with this species his *Isopteron opatroides* exactly agrees, only the latter has clear ferruginous antennæ. The same author's *Platynotus insularis* is, I believe, another member of the genus. The descriptions of these insects and some others, in the 'Transactions of the Entomological Society' (ser. 1. vol. iv.), were very concise; and they were left unticketed, as Prof. Westwood informs me; so that they had afterwards to be determined by these descriptions. As the vast collection of Mr. Hope was at his death in some disorder, it is not impossible that in some instances the true types may have been overlooked.

Cestrinus aversus.

C. elongatus, subdepressus, fuscus, subnitidus, subtiliter sparse griseo-pubescens; elytris striato-punctatis, obovatis.

Hab. Tasmania.

Elongate, subdepressed, dark brown, slightly nitid, finely

---

* M. C. G. Thomson characterizes them as " ovato-globosæ " (Skand. Col. vi. p. 326).

† Wiegm. Arch. 1842, i. p. 172.

and remotely pubescent, the pubescence composed of very small stiff greyish bristles; head closely punctured, clypeus separated from the front by a slightly arched, deeply impressed groove; prothorax a little broader than long, closely punctured, widely emarginate at the apex, the sides slightly rounded and obsoletely crenated, the base truncate; scutellum small, transverse; elytra broader than the prothorax at the base, the greatest breadth towards the apex, striate-punctate, the punctures approximate and deeply impressed; body beneath and femora pitchy brown, finely punctured, tibiæ paler; antennæ and tarsi ferruginous. Length 3 lines.

Cestrinus punctatissimus.

C. elongatus, subdepressus, rufo-fuscus, opacus, subtiliter griseo-pubescens; elytris striatis, creberrime punctatis, lateribus parallelis.

Hab. Tasmania.

Elongate, subdepressed, reddish brown, opaque, with scattered greyish bristles; head and prothorax as in the last; scutellum curvilinearly triangular; elytra broader than the prothorax, the sides nearly parallel, striated, each of the striæ filled with two or three rows of closely impressed irregular punctures; body beneath, legs, and antennæ pale reddish ferruginous, the former and femora punctured. Length 3 lines.

The closely arranged punctures on the elytra, many of them impinging on the lines between the striæ, will readily distinguish this species from the former.

Cestrinus posticus.

C. elongatus, fuscus, subtiliter sparse griseo-pubescens; elytris striato-punctatis, apicem versus elevatis.

Hab. Tasmania.

Elongate, subdepressed, dark brown, with scattered greyish bristles; head and prothorax as in C. *acersus*, but narrower, and the punctures smaller; scutellum confounded with the elytra; the latter gradually broader behind, and, towards the apex, prominently raised at the suture, striato-punctate, the punctures large, squarish, and regularly arranged; body beneath pitchy, finely punctured; legs and antennæ paler. Length $2\frac{3}{4}$ lines.

I have only a single specimen of this insect; but the peculiar elevation of the elytra posteriorly seems to mark it out as a good species.

Nearly related to *Cestrinus* is *Asida serricollis*, Hope[*]; it differs generically in the epipleuræ of its elytra being broader and horizontal or subhorizontal, and the mesosternum entire anteriorly, the last joint of the labial palpi oblong-ovate and somewhat acuminate, and the mentum trapeziform. I propose to call this genus *Achora*.   *Opatrum denticolle*, Blanch.[†], is probably another species.

TYPHOBIA.

Subfamily *DIAPERINÆ*.

Antennæ art. omnibus obconicis, ultimo excepto.
Tarsi postici art. primo elongato.

The character of the antennæ at once separates this genus from *Diaperis*; to this may be added the peculiarly deep opacity of the coloration and the more flattened form. There is a slight transverse elevation on the forehead of one of my specimens[‡].

Typhobia fuliginea.

A. ovalis, subdepressa, nigra, opaca; corpore infra, antennis pedibusque rufo-testaceis, nitidis.

Hab. Queensland; Victoria.

Rather narrowly oval, subdepressed, black, opaque; head somewhat pitchy, finely punctured; prothorax impunctate, anterior angles slightly produced, the lateral marginal line glossy reddish testaceous; scutellum transversely triangular; elytra finely striate-punctate, the punctures minute, the intervals of the striæ broad and very slightly convex; body beneath, legs, and antennæ glossy reddish testaceous. Length $2\frac{1}{4}$ lines.

Platydema§ *aries.*

P. ovalis, modice convexa, nigra, nitida; elytris striato-punctatis, fasciis duabus, ad suturam interruptis, luteis.

Hab. Brisbane.

Oval, moderately convex, black, shining; head finely and

---

[*] Trans. Ent. Soc. ser. 1. iv. p. 108.

[†] Voy. au Pôle Sud, Ins. Col. pl. 10. fig. 13.

[‡] The males of a great many species of the subfamily, especially in the genera *Platydema* and *Arrhenoplita*, have the head furnished with two short horns, either between the eyes or a little above them. But in a species from Brazil, lately given me by Alexander Fry, Esq., these horns are transferred, so to say, to the apex of the prothorax. This remarkable insect will form a new genus. I have adopted the name *Arrhenoplita* of Kirby (Faun. Bor.-Amer. Ins. p. 235) instead of *Hoplocephala*, which had been used years previously by Cuvier for a genus of Ophidians.

§ De Cast. et Brullé, Ann. d. Sci. Nat. xxiii. p. 350.

rather closely punctured, on the inner side and a little above each eye, in the male, a short vertically compressed horn, obliquely truncate at the apex and densely fringed with short yellowish hairs; prothorax twice as broad at the base as long, finely punctured, an oblong fovea on each side posteriorly; scutellum curvilinearly triangular; elytra more convex behind the middle, striate-punctate, the striæ very shallow, the intervals between them broad, flat, and minutely punctured; near the base a broad yellow band, and a similar one near the apex, both interrupted at the suture; body beneath, legs, and antennæ dull luteous, the former clouded with brown. Length 2¼ lines.

Resembles *P. tetraspilota*, Hope, in coloration, but a vastly more bulky insect, and remarkable for the form of the horn, with which the male only is armed.

Platydema oritica.

P. ovalis, modice convexa, nigra, nitida; elytris striato-punctatis; antennis pedibusque pallide ferrugineis.

Hab. Victoria?

More broadly oval than the last, glossy black; head of the male with two horizontal triangular and acuminate horns, tipped with ferruginous, between the eyes; prothorax as in the last, but narrower and more convex; scutellum curvilinearly triangular; elytra more convex at the middle, striate-punctate, the intervals between the striæ convex, minutely punctured; body beneath dark glossy brown; legs and antennæ yellowish ferruginous. Length 2½ lines.

Dr. Howitt has not given me the locality of the above, nor of the following, which differs in some degree generically from *Platydema* in that the fourth, fifth, and sixth joints of the antennæ are obconic, and not transverse, although gradually thicker outwards.

Platydema limacella.

P. breviter ovata, nigra, nitida; elytris striato-punctatis, humeris luteis.

Hab. Victoria?

Shortly ovate, moderately convex, black, shining; head of the male with two short pointed horns, antennary ridges, apex of the clypeus, and antennæ luteous; prothorax finely punctured, twice as broad as long at the base, a little depressed near the scutellum, the margins luteous; scutellum curvilinearly triangular; elytra striate-punctate, the intervals of the striæ minutely punctured, broad, and convex, the shoulders luteous; epipleuræ of the elytra and body beneath dull luteous; legs clear luteous.   Length 2 lines.

The following is, no doubt, a *Platydema*; but there is no trace of horns in either of my two specimens: probably they are both females.

Platydema thallioides.

P. elliptica, convexa, rufo-testacea, nitida; prothorace utrinque macula arcuata, elytrisque (sing.) maculis tribus magnis nigris; antennis basi exceptis nigris.

Hab. Sydney.

Elliptic, convex, reddish testaceous, shining; head finely punctured, rather depressed between the antennary ridges; prothorax smooth, slightly expanded at the lateral margins, a large black arched spot or stripe extending from the anterior to the posterior angles on each side, leaving in the middle of the disk a nearly triangular patch; scutellum curvilinearly triangular; elytra minutely seriate-punctate, on each a round black scutellar spot, and two transverse, also black, the first in the middle, the second near the apex, both large and approaching the suture; body beneath brownish testaceous, the metasternum clouded with black; legs testaceous; antennæ black, the two basal and base of the third joint fulvous testaceous.   Length 2½ lines.

Ceropria?* valga.

C. breviter ovalis, nigra, subnitida, antennis art. duobus basalibus, labro tarsisque fulvis; tibiis intermediis et posticis valde curvatis.

Hab. Queensland.

Shortly oval, black, subnitid, the two basal joints of the antennæ, upper lip, and tarsi fulvous; head very short in front; the clypeus broad, truncate anteriorly, the antennary ridges impinging only slightly on the eyes; antennæ with the fourth and following joints to the tenth inclusive more or less obconic, and only slightly dilated on one side, the last ovate; prothorax nearly twice as broad as long, widely emarginate at the apex, nearly impunctate; scutellum transversely triangular; elytra rather broader than the prothorax, the sides nearly parallel, striate-punctate, the intervals of the striæ flattish; body beneath dark glossy brown; femora and tibiæ pitchy, tibiæ curved, especially the intermediate and posterior.   Length 4 lines.

Differs from *Ceropria* in the antennæ, which are scarcely

* De Cast. et Brullé, Ann. des Sci. Nat. xxiii. p. 396.

serrated on the inner edge, and in the shortness of the head anteriorly, the eyes nearly free, &c. As the genus has a very extended geographical range, and there are only two described species from Australia, it seems best for the present to consider this one an aberrant member.

Pterohelæus* nitidissimus.

Pterohelæus striato-punctatus, De Brême, Essai &c. p. 31, pl. 2. fig. 6 (*nec* Boisduval).

P. ovalis, nitidissime niger; elytris subtilissime seriatim punctatis.

Hab. South Australia.

Oval, moderately convex, very glossy deep black; head finely and closely punctured, clypeal groove broad and shallow; prothorax very minutely and rather closely punctured, rounded at the sides, the edge of the expanded margin anteriorly recurved, an irregular well-marked groove at the base interrupted in the middle; scutellum curvilinearly triangular; elytra a little contracted behind the shoulders, very finely seriate-punctate, the punctures less regularly arranged near the suture; body beneath and legs very glossy black, propectus opaque, granulate; antennæ reaching to the base of the prothorax, third joint half as long again as the fourth. Length 5–5½ lines.

A typical specimen, I believe, in the Oxford Museum shows that this is *P. striato-punctatus*, De Brême; and his description, with one exception, fairly enough accords with it; I hold, however, that it cannot be the same species as that described (?) by Dr. Boisduval in the following words:—" Elongato-ovata nigra; thorace lævi; elytris elongatis, punctis majoribus impressis striatim digestis"†. The exception alluded to is the phrase "fortement ponctué," which may be a slip of the pen for "faiblement ponctué." Dr. Boisduval's "striatim" might in the same way have been intended for "seriatim," but for the specific name "*striato-punctatus*" and the French translation "alignés en stries." There are no striæ whatever in the species before us, nor are there any mentioned by M. de Brême.

Pterohelæus vicarius.

P. sat late ovalis, fusco-niger, nitidus; sulco clypeali distincto; elytris leviter seriatim punctatis.

Hab. Queensland; New South Wales; Victoria.

---

* De Brême, Essai &c. p. 27.
† Voy. de l'Astrol. p. 206.

Rather broadly oval, brownish black, shining; head thickly and roughly punctured, clypeal groove well defined, narrowly and sharply limited, the transverse portion above curved downwards; prothorax minutely but not very closely punctured, rounded at the sides, the expanded margins not recurved, the irregular basal groove on each side nearly obsolete; scutellum broadly triangular, its apex rounded; elytra a little contracted behind the shoulders, finely, but not minutely, seriate-punctate, the punctures less regularly arranged near the suture and base; body beneath and legs glossy brownish black, the propectus opaque, granulate; antennæ short, the third joint nearly twice as long as the fourth.   Length 6–7 lines.

Broader and much less finely punctured than the last species, and not particularly glossy, &c. In some collections it is labelled *P. striato-punctatus*, Boisd.; but the same objection applies to this as to *P. nitidissimus*. Both species have the abdominal segments finely striated longitudinally (a character common to many Tenebrionidæ) and the clypeal grooves well marked.

Pterohelæus litigiosus.

P. paulo anguste ovalis, ferrugineo-fuscus, nitidus; clypeo antice late emarginato, sutura indistincta; elytris tenuiter striato-punctatis.

Hab. Sydney.

Rather narrowly oval, rusty-brown, shining; head finely punctured, a little concave in front; clypeus broadly emarginate anteriorly, separated from the front by a narrow indistinct line; prothorax very minutely punctured, a short longitudinal groove near the apex, none at the base, the expanded margins not recurved; scutellum transversely triangular; elytra callous at the base, rather finely seriate-punctate, the intervals of the rows slightly raised, the fourth and eighth intervals rather more so than the others, the expanded margins narrow; body beneath, legs, antennæ, and margins of the prothorax and elytra reddish ferruginous. Length 7 lines.

In colour and outline resembling *P. silphoides*, but rather broader, and not dull brown as in that species, the intervals of the striæ more elevated, the punctures larger, and, above all, a broad callosity at the base of the elytra.

Pterohelæus alternatus.

P. subanguste ovalis, niger, nitidus; clypeo antice vix emarginato, sutura fere obsoleta; elytris in medio planatis, leviter seriatim punctatis, interstitiis alternis elevatis.

Hab. " Interior."

Rather narrowly oval, black, shining, somewhat depressed; head finely punctured; clypeus scarcely emarginate in front, its suture nearly obsolete; prothorax minutely punctured, a broad shallow fovea on each side at the base, no groove, the expanded margins not recurved; scutellum curvilinearly triangular; elytra flattish at the middle and base, finely seriate-punctate, the alternate intervals of the rows raised, the fourth, eighth, twelfth, and sixteenth (the last) much more so than the others, the expanded margins broad at the base, gradually narrower to the apex; body beneath and legs black, slightly glossy, tibiæ covered with short spinous hairs; antennæ short, not reaching to the end of the prothorax, black. Length 8 lines.

A very distinct species, in outline resembling *P. Reichei,* but the elytra with expanded margins and strongly marked elevated lines, &c. Dr. Howitt merely gives " Interior " as its locality.

Pterohelæus minimus.

P. oblongo-ovalis, piceus, subnitidus, marginibus clypeoque pallidioribus; prothorace confertissime oblongo-punctato; elytris subtuberculatis, subtiliter et vage punctatis.

Hab. Cooper's Creek.

Oblong-oval, pitchy brown, subnitid, the margins of the prothorax and elytra, and the anterior part of the head paler, yellowish brown; head densely punctured, the clypeal groove very indistinct; prothorax rather short, covered with fine oblong punctures, the intervals very narrow, and in certain lights causing the surface to assume a delicately corrugated appearance, the expanded margins narrow and slightly reflected; scutellum transversely triangular; elytra minutely and irregularly punctured, with scattered minute tubercles, especially near the suture, the expanded margins very narrow; body beneath and legs glossy reddish testaceous; antennæ short, inclining to testaceous. Length 3½ lines.

The smallest species of the genus, and very distinct on account of the sculpture of the prothorax and elytra. I have placed it after *Pterohelæus peltatus,* Er., which it resembles in outline.

The three following are closely allied in general appearance,

but are distinguished by several small but well-marked points
of difference. They seem to lie between *P. Walkerii* and *P.
silphoides*, not so broad as the first nor so narrow as the last,
and all moderately convex. Two of these species have the
sutural margin raised; one (*P. laticollis*) has the expanded
margins of the elytra rather broad, the broadest part in the
middle; the other (*P. hepaticus*) has them much narrower,
very slightly contracted behind the shoulders, the rest to beyond the middle of nearly equal breadth; the third (*P. dispar*)
affects two forms, apparently depending upon sex, the male
being elliptic, the female obovate; in this the sutural margin
is without any elevation.

Pterohelæus laticollis.

P. fuscus, nitidus, marginibus dilutioribus; oculis approximatis;
prothorace elytris latiore, his postice gradatim angustioribus.

Hab. Melbourne.

Dark glossy brown, the expanded margins of the elytra and
prothorax considerably paler; head rather narrow behind the
antennary ridges, concave between them; the eyes rather large
and approximate; clypeus very convex, except at its anterior
angles, its suture indistinct; prothorax short, broader than the
elytra at its base, minutely punctured, the margins broad and
only very slightly reflected, the basal foveæ strongly impressed;
scutellum curvilinearly triangular; elytra gradually and rather
rapidly narrowing from the base, seriate-punctate, the alternate
intervals of the rows forming slightly elevated lines, the suture strongly elevated from below the scutellar striola, the
punctures rather small, the expanded margins, owing to a
contraction of the sides of the disk, broadest at the middle,
behind very distinctly reflected; body beneath and femora
very glossy chestnut-brown; antennæ, tibiæ, tarsi, and epipleuræ of the elytra reddish ferruginous. Length 10 lines.

Pterohelæus hepaticus.

P. fuscus (aliquando rufo-brunneus), subnitidus, marginibus dilutioribus; oculis distantibus; prothorace elytris haud latiore, his
postice gradatim angustioribus.

Hab. Melbourne.

Dark brown (or sometimes light reddish brown), paler at
the margins, less glossy than the last; head rather narrow
behind the antennary ridges; the clypeus very convex, its suture above indistinct, but forming a well-marked groove on

each side; the eyes widely apart; prothorax not broader than the elytra at their base, much longer and narrower than in the last, the basal foveæ represented by a large shallow depression on each side; scutellum transversely triangular, the sides curvilinear; elytra gradually narrowing from the base, the sides of the disk not contracted, seriate-punctate, the intervals of the rows not raised, the punctures rather small, the expanded margins of nearly equal breadth, or only very gradually narrowing behind, the suture raised as in the last; body beneath and legs glossy chestnut-brown; antennæ glossy ferruginous. Length 8½ lines.

Pterohelæus dispar.

P. breviter ellipticus (♂), oblongo-obovatus (♀), piceus, nitidus, marginibus dilutioribus; oculis haud distantibus; elytris basin versus parallelis (♂ in medio paulo latioribus), lineis elevatis nullis.

Hab. Swan River.

Shortly elliptic in the male, oblong-obovate in the female, shining pitchy brown, the margins much paler; head rather narrow in front; clypeus convex, its suture rather indistinct; the eyes not remote; prothorax shorter proportionally in the male, the basal foveæ shallow, between them opposite to the scutellum an indistinct groove; scutellum triangular; elytra nearly parallel at the sides, and not broader than the prothorax in the female, broader in the middle in the male, finely seriate-punctate, the intervals without raised lines, the suture not elevated, the expanded margins of nearly equal breadth at the sides, and a little reflected at the edge; body beneath and femora dark chestnut-brown, shining; antennæ, tibiæ, and tarsi paler. Length (♂) 7, (♀) 9 lines.

Helæus squamosus (Howitt's MS.).   Pl. XII. f. 4.

H. oblongus, parallelus, ferrugineo-fuscus, opacus, sparse fulvo-squamosus; elytris sing. unicostatis.

Hab. Cooper's Creek; Darling River.

Oblong, parallel at the sides, impunctate, rusty-brown, opaque, sparsely covered with fulvous hairs simulating scales; head a little prolonged anteriorly; clypeus rounded; prothorax rather transverse, with a strongly marked carina in the middle, the foliaceous margins broad and reflexed; scutellum transversely triangular; elytra moderately convex, depressed along the sutural region, the suture finely raised, and near it on each side a strongly marked carina, which terminates abruptly at a little distance from the apex, a line of small tubercles towards the foliaceous margins, which are moderately broad, but expanded inwardly near the shoulders; body beneath and legs opaque rusty-brown clothed with fine scattered hairs. Length 12 lines.

A very distinct species, having no similitude to any of its congeners. Unfortunately, it is not quite perfect as to its antennæ and anterior tarsi; and their reproduction in the figure must be taken with a slight reservation. In fresh examples, it is very likely the flattened hairs (they are not true scales) are more numerous than I have represented.

Saragus limbatus.

S. late ovalis, modice convexus, nigrescens vix nitidus; elytris leviter seriatim punctatis, interstitiis alternis paulo elevatis, latera versus sensim minus conspicuis.

Hab. Melbourne; Gawler.

Broadly oval, moderately convex, brownish black, scarcely nitid; head and prothorax finely punctured, the latter slightly convex, the basal foveæ nearly obsolete, the anterior angles rounded, posterior produced and recurved, foliaceous margins moderately broad, a little reflexed, and edged with a thickened border; scutellum transversely triangular; elytra not broader than the prothorax, finely seriate-punctate, the intermediate spaces between the rows raised, three or four on each side the suture the most so, those towards the sides gradually disappearing, foliaceous margins narrowing gradually posteriorly, transversely corrugated; body beneath and legs dark chestnut-brown, a little glossy, the abdominal segments longitudinally corrugated; antennæ ferruginous brown. Length 7 lines.

In outline resembling *S. simplex*, Hope (=*S. asidoides*, Pasc.), but differing in the sculpture of the elytra, &c. Dr. Howitt sends me another *Saragus*, from Port Augusta in South Australia, unfortunately without head or legs, but certainly one of the most remarkable of the subfamily. *S. australis*, Bois., seems to be not the same described under that name by the Marquis de Brême.

Dr. Howitt has sent me not less than four new genera of that handsome and almost exclusively Australian* subfamily, Cyphaleinæ. As a considerable addition has now been made

---

* The only exception is a Sumatran insect, which I have recently characterized under the name of *Artactes nigritarsis* (Proc. Ent. Soc. 1868, p. xii). It will be more fully described and figured hereafter.

to the group since M. Lacordaire's volume was published in 1859, the following tabulation may be useful:—

Prosternum prolonged and compressed anteriorly (carinated).
  Antennæ rather short, joints gradually thicker and shorter from 7th or 8th to 10th.
    Tibiæ dilated at the end .... *Lepispilus,* Westw.
    Tibiæ not dilated.
      Body glabrous.
        Epipleuræ of the elytra entire.
          Intercoxal process broad, slightly rounded anteriorly .... *Platyphanes,* Westw.
          Intercoxal process narrower and triangular.
            Antennæ with the three penultimate joints obconic .... *Hectus,* n. g.
            Antennæ with the two penultimate joints transverse .... *Opigenia,* n. g.
        Epipleuræ of the elytra incomplete or suddenly narrowed behind.
        Body oblong, depressed.
          Antennæ with the penultimate joints oblong .... *Olisthæna,* Er.
          Antennæ with the penultimate joints transversely obconic .... *Decialma,* n. g.
        Body hemispherical .... *Hemicyclus,* Westw.
      Body pilose .... *Altes,* n. g.
  Antennæ rather long, the penultimate joints little thicker than the rest.
    Basal joint of posterior tarsi as long as the rest together .... *Chartopteryx,* Westw.
    Basal joint of posterior tarsi shorter .... *Oremasis,* Pasc.
Prosternum not prolonged or compressed anteriorly.
  Mesosternum notched for the reception of prosternal process.
    Tarsi pilose beneath.
      Body oblong.
        Epipleuræ of the elytra suddenly narrowed behind .... *Prophanes,* Westw.
        Epipleuræ of the elytra gradually narrowed behind.
          Eyes partially covered by the prothorax *Lygestira,* Pasc.
          Eyes clear of the prothorax .... *Cyphaleus,* Westw.
      Body hemispherical .... *Artactes,* Pasc.
    Tarsi partially pilose beneath .... *Barytipha,* n. g.
  Mesosternum not notched .... *Mithippia,* n. g.

OPIGENIA.

Subfamily CYPHALEINÆ.

Oculi liberi.
Antennæ breviusculæ, art. 9°, 10° transverse obconicis.
Mesosternum breve, profunde incisum.

    Head not inserted in the prothorax so far as the eyes; clypeus truncate in front, its suture obsolete. Eyes moderate, distant above. Antennæ rather short, the third joint twice as long as the second, fourth to eighth gradually shorter and broader, ninth and tenth transversely obconical, the last rounded. Mentum trapeziform, narrow at the base, strongly convex on the median line; lower lip transverse, rounded at the sides, slightly emarginate in front. Maxillary lobes narrow, the inner falciform, not produced into a hook. Labial palpi with the last joint very large, broadly obconic. Prothorax transverse, broadly emarginate at the apex, anterior angles rounded. Elytra oblong, convex; epipleuræ gradually narrower behind. Legs rather short; basal joint of the posterior tarsi longer than the two next together. Mesosternum deeply notched. Intercoxal process narrowly triangular.

The type of this genus has no very obvious affinity to, and is different in habit from all others of this subfamily, although its technical characters are not very special. The internal maxillary lobe, unlike most of the genera of the Cyphaleinæ, is not produced into a hook, although the apex is pointed.

Opigenia iridescens.

O. oblongo-ovata, modice convexa, aureo-viridis, in certo situ purpureo resplendens.

Hab. Victoria.

Oblong-ovate, moderately convex, golden-green, with rich purple reflections; head rather finely and closely punctured; antennæ glossy ferruginous; prothorax finely but less closely punctured than the head; scutellum triangular, black; elytra broader than the prothorax, their greatest breadth a little behind the middle, seriate-punctate, the punctures small and not approximate, the intervals of the rows broad and finely punctured; body beneath and legs glossy black, the former finely punctured. Length 6 lines.

Hectus.

Subfamily Cyphaleinæ.

Prosternum antice productum, carinatum.
Processus intercoxalis brevis, antice rotundatus.
Oculi liberi.

In other respects this genus agrees with *Lygestira*, except that it has no raised lines on the elytra—if that be a generic character. My specimen, the only one I have seen, appears to be a male, but it has the anterior tarsi only dilated; in *Lygestira*, judging from the few examples I have been able to examine, the intermediate tarsi as well are dilated, although but slightly.

Heetus anthracinus. Pl. XII. fig. 6.

H. modice convexus, æneo-niger, nitidissimus; elytris vage et sparse punctatis.

Hab. Rockhampton.

Moderately convex, not depressed, brassy black, very glossy; head and prothorax finely punctured, anterior angles of the latter strongly produced and acuminate; scutellum nearly equilaterally triangular; elytra a little broader than the prothorax at the base, their sides slightly parallel, not broader behind, sparingly and irregularly punctured, the punctures of moderate size; body beneath and legs brownish black, very glossy; antennæ dark ferruginous. Length 6 lines.

Lepispilus Stygianus.*

L. niger, nitidus; prothorace brevi, valde transverso, angulis anticis haud producto, rotundato.

Hab. "Alps of Victoria."

Entirely black, glabrous, shining; head small comparatively; clypeus not distinctly separated from the front, its punctures not more crowded than those on the rest of the head; prothorax short, very transverse, minutely punctured, anterior angles not produced, broadly rounded; scutellum equilaterally triangular; elytra large, very convex, much broader behind (probably in ♀ only), with rather fine punctures irregularly crowded, and here and there almost obliterated, with no traces of lines or foveated impressions; body beneath and legs glabrous and glossy, the tibiæ thickly punctured and strongly dilated at the tips. Length 10 lines.

Radically distinct from its only congener (*L. sulcicollis,* Hope) in its colour, sculpture, absence of pubescence, and form of prothorax. My specimen appears to be a female.

ALTES.

Subfamily *CYPHALEINÆ.*

Corpus longe pilosum.
Antennæ breves, art. duobus penultimis transversis.
Tibiæ lineares, ant. et interm. haud calcaratæ.
Tarsi postici art. basali breviusculo.

These characters separate this genus from *Chætopteryx,* Westw., to which I had doubtfully referred the species (*C. binodosa*) constituting its type. It is perhaps the most remarkable of all the Cyphaleinæ, on account of the large hump at the base of each elytron, precisely as in the Brazilian genera

* Westwood, Arcan. Ent. i. 44.

Dicyrtus and *Thecacerus*. *Altes binodosus*, represented on Plate XII. fig. 2, is an ovate, convex, dark copper-brown insect, sparsely furnished with long flying hairs on the body and legs. *Chartopteryx* has erect scale-like hairs, rather thickly clustering at the base of the elytra, very different in their character and distribution from those on *Altes*.

DECIALMA.

Subfamily CYPHALEINÆ.

Antennæ art. penultimis breviter obconicis.
Tibiæ obsolete calcaratæ.
Tarsi lineares, art. ult. elongato.

Head exserted; clypeus broad, separated from the front by a straight groove. Eyes not contiguous to the prothorax, prominent, broad, nearly entire. Antennæ short, slender, the last six joints thicker than the rest, third shorter than the two next together, all, except the last, more or less obconic, the last ovate. Mentum trapezoidal, narrow at the base. Prothorax transverse, broadly emarginate at the apex, slightly foliaceous at the sides. Elytra oblong, slightly depressed; epipleuræ obliquely descending, nearly obsolete towards the apex. Legs short; femora thickened; tibiæ linear, very shortly spurred; tarsi slender, the claw-joint elongate. Prosternum produced behind. Mesosternum with a V-shaped notch.

It is with some hesitation that I propose this as a genus distinct from *Olisthæna*, Er.*, which is unknown to me, but with which it agrees, so far as he has characterized it, with the exception of the antennæ: these he describes as having the penultimate joints longer than they are broad, by which character he differentiates it from *Pachyculia* (= *Lepispilus*). On the contrary, in *Decialma* the penultimate joints are broader than they are long; and in a subfamily like the Cyphaleinæ, remarkable for a difference of habit without a correlative difference of structure, a character like the above becomes of importance.

Decialma tenuitarsis.

D. oblonga, modice convexa, nitida; capite prothoraceque nigris; elytris fuscis, vage punctatis.

Hab. Victoria.

Oblong, moderately convex, shining; head black, very closely and rather finely punctured, but with few punctures on the clypeus; prothorax black, minutely and sparsely punc-

* Wiegm. Arch. 1842, Bd. i. p. 177.

tured, very short, the sides nearly parallel, but a little rounded anteriorly, anterior angles slightly produced; scutellum brown, curvilinearly triangular; elytra a little broader than the prothorax, parallel at the sides, irregularly covered with small approximate punctures; body beneath and legs glossy chestnut-brown, with minute scattered punctures; antennæ not reaching to the base of the prothorax, and, with the tarsi, dull glossy ferruginous. Length 5 lines.

BARYTIPHA.

Subfamily CYPHALEINÆ.

Antennæ breviusculæ, art. 8°, 9°, 10° transversis.
Epipleuræ elytrorum postice vix angustiores.
Tarsi subtus apice breviter pilosi.

Head deeply inserted in the prothorax, convex in front; clypeus strongly emarginate, its groove arched. Eyes narrow, transverse, constricted in the middle, distant above. Antennæ rather short, third joint twice as long as the second, fourth to seventh gradually shorter, eighth, ninth, and tenth transverse, the last rounded. Mentum broadly subcordiform, its face concave; lower lip rounded anteriorly. Maxillæ short, the inner lobe strongly hooked. Maxillary palpi securiform, labial subobconic. Prothorax transverse, apex broadly emarginate, anterior angles not produced. Elytra slightly broader than the prothorax at the base, their sides subparallel; epipleuræ, except at the base, nearly equal in width throughout. Legs rather short; tibiæ gradually broader below; tarsi shortly pilose at the apex, basal joint of the posterior not longer than the two next together. Metasternum rather short; interfemoral process narrowly triangular.

The peculiar vestiture of the tarsi (composed of short stiff hairs confined to the apices of the joints) is exceptional, and at once differentiates this genus. Dr. Howitt tells me that the species described below is gregarious in old deserted swallows' nests in hollow and decaying trees.

Barytipha socialis. Pl. XII. fig. 5.

B. fusca (aliquando brunnea), subnitida; elytris fere opacis, subtiliter substriato-punctatis.

Hab. Victoria.

Dark brown, sometimes reddish brown; head and prothorax subnitid, very minutely punctured, the latter regularly but not very convex above; scutellum rather broadly triangular; elytra somewhat opaque, lightly striate-punctate, the punctures minute, approximate, the intervals of the striæ slightly convex, the alternate ones rather more raised; body beneath brownish, the abdomen marked with delicate longitudinal lines; antennæ and tarsi ferruginous, shining.   Length 7 lines.

Mithippia.

Subfamily *Cyphaleinæ.*

Oculi prothorace haud liberi.
Antennæ art. haud transversis, tribus ult. gradatim crassioribus.
Mesosternum amplum, declive, haud excisum.

Head deeply inserted, rounded anteriorly; clypeus separated from the front by a shallow groove. Eyes partly covered by the prothorax, transverse, broad, remote. Antennæ slender, none of the joints transverse, the last three a little stouter than the rest. Mentum trapezoidal, narrowed at the base. Prothorax subquadrate, flattish, broadly emarginate at the apex, with the anterior angles produced, the sides forming a narrow carina. Elytra oblong, slightly depressed; epipleuræ obliquely descending, entire. Legs rather short; femora slightly thickened; tibiæ linear; tarsi with the basal joint of the intermediate and posterior elongate, the last joint of all short. Prosternum rather broad, depressed, not produced behind. Mesosternum large, declivous, not notched.

A degraded form of the Cyphaleinæ, differing from the rest in its simple mesosternum, not notched for the reception of the prosternal process; the mesosternum, notwithstanding, preserves the peculiar hollowed surface which forms one of the characteristics of the subfamily.

Mithippia aurita.   Pl. XII. fig. 3.

M. oblonga, depressa, brunnea, subnitida.

Hab. Adelaide.

Oblong, depressed, clear reddish brown, somewhat nitid; head and prothorax very closely covered with oblong, rather small but deep punctures; the latter subquadrate, slightly rounded at the sides anteriorly, but a little incurved behind the middle, a shallow transverse impression towards the base; scutellum semicircular; elytra very closely striate-punctate, the punctures large, square, and placed nearly at equal distances both transversely and longitudinally, and each giving rise to a single recurved hair; body beneath brownish testaceous, shining, with rather crowded piliferous punctures; femora and tibiæ darker, closely punctured; tarsi and antennæ yellowish ferruginous.   Length 4¾ lines.

Achthosus* laticornis.

A. fusco-castaneus, nitidus; clypeo haud cornuto; prothorace (♂)
in medio apicis leviter excavato; antennis art. 6 penultimis valde
transversis.

Hab. Clarence River.

Dark chestnut-brown, glossy, slightly convex; head small,
a broad triangular excavation between the eyes; clypeus very
convex, not horned; antennæ reddish, gradually broader to
the seventh joint, the six penultimate very transverse; prothorax broader than long, moderately convex, slightly rounded
at the sides, finely punctured, the middle of the apex with
an irregular excavation; scutellum small, triangular; elytra
deeply striate-punctate, the punctures rather small and not
approximate; body beneath reddish brown, glossy; legs paler;
anterior tibiæ dilated, serrated externally, and emarginate internally near the base; middle tibiæ rather spined than serrated.   Length 5½ lines.

The female differs in the prothorax being without any excavation, the anterior tibiæ without the internal emargination,
and the somewhat smaller size.   The type of the genus is a
much larger and almost cylindrical insect, with a deep excavation occupying nearly the whole anterior portion of the prothorax, and with a short broad horn on the clypeus.   I have
another species, from New Zealand†, closely resembling the
above, but, from its simple prothorax, a lowering of the type.
This genus is represented in South America by *Antimachus*,
which also includes similarly degraded forms.

TYNDARISUS.

Subfamily STRONGYLIINÆ.

Antennæ breviusculæ, ad apicem sensim crassiores, art. ultimo præ-
cedente duplo longiore.
Prothorax transversus, lateribus marginato-productus.
Tarsi longissimi, lineares, omnes æquales.

Head small, subvertical, narrower anteriorly; clypeus truncate at the apex; labrum prominent.   Eyes broad, vertical,
not approximate.   Antennæ rather short, a little thicker outwards, the third joint longer than the fourth, the last oval,
twice the length of the preceding.   Mentum trapeziform;
lower lip as large as the mentum, rounded in front, slightly
emarginate at the apex; its palpi stout, with the last joint
large and subsecuriform.   Maxillæ small, outer lobe transverse,

---

* Pascoe, Journ. of Entom. ii. p. 42.   It belongs to the Ulominæ.
† Probably *Uloma lævicostata*, Blanch.

strongly fringed, inner lobe narrow, elongate, and unarmed; the palpi with the last joint narrowly securiform. Prothorax small, transverse, a little expanded at the sides, the pronotum separated from the flanks by a well-marked carina. Elytra very ample, oblong, convex, slightly incurved at the sides; epipleuræ entire and channelled nearly throughout. Legs slender; femora rather short, fusiform; tibiæ thicker below, manifestly spurred, the posterior longest; tarsi slender, as long as or longer than their tibiæ, the anterior as long as the intermediate and posterior, thickly pilose beneath. Prosternum elevated, a little produced behind; mesosternum V-shaped. Intercoxal process triangular.

The state of the subfamily to which this genus belongs is at present one of the most unsatisfactory of all the Heteromera. The typical genus *Strongylium**, which has been recently elaborately monographed by M. Mäklin†, contains 266 species, exclusive of those in English collections; and, as may be supposed, there is no more definite generic idea to be obtained from such a number than there would be from the same number in any one of the so-called genera of the Linnean epoch. Putting, therefore, *Strongylium* aside as merely a designation for a collective number of discrepant forms, the genus before us may be at once distinguished from all others of the subfamily by the great length of the anterior tarsi, which if anything rather exceed the rest in that respect. The prothorax is also very different from anything that obtains in the other genera of this group, except *Dicyrtus* and *Psydus*. I am unable to give the sex of my specimen, or to say if there are any sexual differences. Dr. Howitt has not given me its exact habitat.

Tyndarisus longitarsis. Pl. XII. fig. 1.

T. cupreo-brunneus, nitidus; elytris substriato-punctatis.

Hab. Australia.

Copper-brown, glossy; head distinctly and closely punctured; clypeus imperfectly separated from the front; antennæ extending a little beyond the prothorax, joints five to ten gradually thicker and shorter, of a paler colour, and pubescent; prothorax finely punctured, almost twice as broad as long, rounded at the sides anteriorly, a little incurved behind the middle, with the posterior angles acuminate, the apex slightly emarginate, the base with a broad middle lobe; scutellum curvilinearly triangular, the middle pilose; elytra much broader than the prothorax, and about five times its length, oblong, a

---

* Established by Kirby, in 1818, in the Trans. Linn. Soc. xii. p. 417.
† Acta Soc. Sci. Fennicæ, viii. p. 117 (1866).

little narrower in the middle, nearly obsoletely striate-punctate, punctures minute, intervals of the striæ feebly raised; body beneath and legs dark brown, glossy, with a thin greyish pubescence. Length 9 lines.

Notwithstanding the following additions to the genus *Amarygmus*[*], there still remains a considerable number of species, some of which, although they would be called "evidently distinct," are apparently so nearly allied to others already published that they could not be satisfactorily differentiated without a larger suite of specimens than I possess. The variability of some of them (e. g. *A. purpureus*) has probably led to more than one being split up into so-called species. All here described are, I venture to think, either more or less specialized or are distinguished by very strong characters from those to which they may be considered most approximate.

Amarygmus cælestis.

A. ovalis, niger, nitidus; elytris læte cyaneis, sat leviter seriatim punctatis, interstitiis impunctatis; tarsis tenuiter elongatis.

Hab. Brisbane.

Moderately oval, black, shining; the elytra bright indigo-blue, with slight violet reflections; head flat between the antennary tubers, separated from the clypeus by a deep broad groove; eyes approximate, entirely concealed by the prothorax in repose; antennæ rather slender, the last joint irregularly oblong-obovate; prothorax small, not very broad at the base or apex, the punctures almost obsolete; scutellum small, triangular; elytra moderately convex, seriate-punctate, the punctures rather fine, but well-marked, intervals of the striæ impunctate; body beneath and legs glossy black, with a brownish tinge; abdomen finely striated longitudinally; all the tarsi slender, elongate. Length 5½ lines.

A handsome species, allied to *A. amethystinus*, Fab.[†]; the latter, however, has a dark-blue prothorax and red femora.

Amarygmus vinosus.

A. ovalis, viridis, nitidus; elytris purpureis in virides mutantibus, sat leviter seriatim punctatis, interstitiis subtiliter punctatis.

Hab. Sydney.

Moderately oval, green, shining, the elytra purple changing

* Dalman, Anal. Entom. p. 60. M. Blessig separates the Australian species of the genus, under the name of *Chalcopterus*, on account of the mandibles of the latter being entire at the end, not bifid. (Hor. Soc. Ent Ross. fasc. i. p. 103.)   † Ent. Syst. ii. p. 40 (*Erotylus*).

to green according to the light; head black, very slightly convex above the clypeus; eyes moderately approximate; antennæ gradually thicker outward, the last joint ovate; prothorax minutely and sparsely punctured, broad at the apex; scutellum small, triangular, black; elytra moderately convex, seriate-punctate, the punctures rather fine, but well marked, the intervals of the rows minutely punctured; body beneath and legs glossy brownish black, the abdomen finely striated longitudinally; anterior and intermediate tarsi shorter than in the last.   Length 5½ lines.

In outline resembling the last, but differently coloured, with the prothorax shorter and much broader at the apex, the intervals of the rows on the elytra minutely punctured, &c.

Amarygmus exilis.

A. anguste oblongus, nitidus; prothorace trapezoidali, viridi-metallico; elytris elongatis, aureo-viridibus, in certo situ cupreo-resplendens, leviter seriatim punctatis.

Hab. Lachlan River.

Narrowly oblong, nitid, slightly convex; head green in front, blackish towards the clypeus; eyes remote; antennæ short, stout, ferruginous, the last six joints thicker and longer than the two preceding, the third only a little longer than the fourth; prothorax metallic green, trapezoidal, the sides nearly straight, the base not much broader than the apex; scutellum equilaterally triangular, black; elytra rather long, but much broader than the prothorax at the base, the sides nearly parallel, finely seriate-punctate, bright golden-green with copper reflections; body beneath light chestnut, glossy; legs dark chestnut; tarsi ferruginous.   Length 3¼ lines.

A much narrower form than any of the preceding, with shorter antennæ.

Amarygmus indigaceus.

A. oblongus, subnitidus; prothorace nigro, angulis anticis acuminatis; elytris cyaneis, distincte seriatim punctatis; antennis tarsisque obscure testaceis.

Hab. Sydney.

Oblong, a little nitid; head black, rather coarsely punctured; eyes somewhat approximate; antennæ dull testaceous, the last four joints shorter and a little thicker than the others; prothorax rather narrow, shining, minutely punctured, the anterior angles produced and pointed; scutellum triangular, black; elytra broadest at the shoulders, very gradually narrower posteriorly, indigo-blue, rather finely seriate-punctate, the intervals of the striæ narrow; body beneath and legs chestnut-brown, slightly glossy; tarsi dull testaceous. Length 3½ lines.

Allied to *A. picicornis* and *A. tarsalis*; the former, *inter alia*, has varying metallic elytra, and the latter a different prothorax and coarsely punctured elytra.

Amarygmus Cupido.

A. oblongo-ovalis, nitidus; prothorace nigro; elytris læte violaceis, certo situ cyaneo-resplendens, leviter seriatim punctatis.

Hab. Queensland.

Oblong-oval, nitid; head black; eyes scarcely approximate; antennæ dark ferruginous, the last five joints thicker and longer than the three preceding, but the third a little longer; prothorax glossy black, rather broad at the base, finely punctured; scutellum equilaterally triangular, black; elytra broadest, with the sides nearly parallel, along the middle third, rich violet, with lightish blue reflections, finely seriate-punctate; body beneath and legs chestnut-brown, glossy; tarsi ferruginous. Length 3½ lines.

This is a beautiful and very distinct species, in size and form resembling the last, but with a nearly perfectly oval outline.

Amarygmus pusillus.

A. ovalis, niger, parum nitidus; elytris fero opacis, striato-punctatis, interstitiis impunctatis; subtus pedibusque castaneis.

Hab. Kiama.

Oval, black, a little nitid; head dull black; eyes not approximate; antennæ brown, gradually thicker from the third joint, the last five especially so; prothorax a little more nitid, well rounded at the sides, very minutely punctured; scutellum triangular, black; elytra nearly opaque, brownish black, strongly striate-punctate, the punctures oblong-linear, the intervals of the striæ broad, flattish, with a slight trace, in certain lights, of transverse linear impressions; body beneath and legs chestnut-brown, slightly glossy. Length 2¾ lines.

The sculpture of the elytra seems to approach in its character that of *A. rugosus*, Germ.; but the latter, *inter alia*, has rugose striæ, which is not the case in the species before us. The following has also striated elytra, but with different sculpture.

Amarygmus minutus.

A. suboblongo-ovalis; elytris fuscis, subnitidis, fortiter striato-punctatis, interstitiis punctulatis; subtus pedibusque pallide ferrugineis.

Hab. Sydney.

Slightly oblong-oval; head dark brown; eyes not approximate; antennæ pale ferruginous, the last five joints gradually thicker and very little longer than the three preceding; prothorax chestnut-brown, finely punctured, well rounded at the sides, broad at the base; scutellum triangular, brown; elytra brownish black, rather glossy, strongly striate-punctate, the punctures large, round, the intervals between the striæ finely punctured; body beneath and legs pale ferruginous. Length $2\frac{1}{4}$ lines.

Amarygmus obtusus.

A. oblongo-subovalis, niger, nitidus; elytris fusco-purpureis, haud versicoloribus, fortiter seriatim punctatis; tarsis subtus longe pilosis.

Hab. Queensland.

Oblong-suboval, black, shining, the elytra dark purple-brown without reflections; head slightly concave above the clypeus; eyes approximate; antennæ short, slender, the last four joints with a brownish pubescence, the last shortly ovate; prothorax short, gradually rounded from the base, broad and obtuse anteriorly, minutely punctured; scutellum triangular, black; elytra rather elongate, convex, the sides very slightly rounded, seriate-punctate, the punctures rather coarse, the intervals of the rows almost obsoletely punctured; body beneath and legs brownish black, subnitid; tarsi slender, with long hairs beneath. Length 7 lines.

The form and colour of this very distinct species will render it easily recognizable.

Amarygmus polychromus.

A. late ovalis, niger, nitidus; elytris cyaneis, vel purpureis, vel viridibus, coloribus variis resplendentibus, tenuissime seriatim punctatis; prothorace basi haud lato.

Hab. South Australia.

Broadly oval, black, shining; the elytra blue, purple, or green, with metallic reflections of various colours; head very slightly convex in front; eyes moderately approximate; antennæ rather slender, last joint elongate-ovate; prothorax black, with greenish reflections, not broad at the base, narrow at the apex, minutely and sparsely punctured; scutellum equilaterally triangular; elytra rather strongly convex, very finely seriate-punctate, the intervals of the rows broad and minutely punctured; body beneath, legs, and antennæ jet-black, glossy, abdomen finely punctured. Length 6–7 lines.

A very variable species as to the colour of its elytra, but readily distinguished, except from the next, by the fineness of its seriated punctures, which are scarcely to be discriminated from the interstitial punctures, together with its greater breadth and convexity. *A. Howittii* is a still broader species, with its dark-green colour varying principally from darker to lighter shades.

Amarygmus Howittii.

A. late ovalis, nitidus; prothorace nigro; elytris subtiliter seriatim punctatis, interstitiis subtilissime punctatis, æneo et cupreo versicoloribus.

Hab. Victoria.

Broadly oval, smooth, shining; head glossy black; eyes not approximate; clypeus rather narrow; antennæ black, extending but little beyond the prothorax, thicker outwards, the last four joints shorter than the preceding, opaque, the rest glossy; prothorax small, broad at the base, the apex narrow, the anterior angles not produced, glossy black, finely and rather remotely punctured; scutellum small, triangular, black; elytra rather strongly convex, the sides nearly parallel or but very slightly rounded, very finely seriate-punctate, the punctures very close, the intervals of the rows wide and very minutely punctured, the colour dark greenish, shaded from brassy to copper according to the light; body beneath and legs jet-black and very glossy. Length 7 lines.

Dr. Howitt says of this very distinct species, "common everywhere;" but the two specimens he has sent me are the only ones I have seen. It approaches the following in outline, but is very different in colour and sculpture.

Amarygmus semiticus.

A. late obovatus, subnitidus, flavo-cupreus; elytris subtiliter seriatim punctatis, interstitiis vage et subtilissime punctatis; corpore subtus viridi-nigro.

Hab. Port Denison.

Broadly obovate, slightly nitid, colour above an unvarying yellowish copper; head black; eyes approximate; clypeus rather narrow; antennæ black, thicker outwards, fourth and succeeding joints of nearly equal length, the third not much longer than the fourth; prothorax much narrower at the apex, anterior angles somewhat produced, very minutely punctured; scutellum small, curvilinearly triangular, black; elytra broadest nearly at the

base, then rounded, gradually narrower to the apex, finely
seriate-punctate, punctures close, the intervals of the rows
wide, sparsely and very minutely punctured; body beneath
glossy greenish black; legs black.   Length 7 lines.

Amarygmus semissis.

A. breviter ovalis, modice convexus, niger, nitidus; antennis art. 4
ultimis tarsisque fulvo-ferrugineis; elytris leviter striato-punctatis.

Hab. Kiama.

Shortly oval, black, nitid, moderately convex; head scarcely
concave in front; eyes moderately approximated; antennæ
slender, the two basal joints glossy ferruginous, the last four
pubescent, tawny ferruginous, opaque; prothorax small, not
broad at the base, minutely punctured; scutellum triangular;
elytra striate-punctate, the striæ shallow, the punctures rather
fine, the interstices of the striæ very minutely and sparsely
punctured; body beneath and legs glossy brownish black,
tarsi tawny ferruginous.   Length 4 lines.

This species is allied to the following, but, *inter alia*, has a
broader and less elliptic outline, and is much less convex.

Amarygmus ellipsoides.

A. breviter elliptico-ovalis, sat fortiter convexus, fusco-niger, nitidus;
elytris viridi-nigris, leviter striato-punctatis.

Hab. Queensland.

Shortly elliptic oval, rather strongly convex, brownish
black, shining; the elytra greenish black, without reflections;
head scarcely concave in front, a little depressed along the
clypeal groove; eyes not approximate; antennæ glossy ferruginous, long, slender, the last joint narrowly oblong; prothorax
small, rather narrow at the apex, very minutely punctured;
scutellum triangular; elytra striate-punctate, the striæ shallow,
the punctures rather fine, the interstices obsoletely punctured;
body beneath and femora glossy brownish black; tarsi slender
and, with the tibiæ, ferruginous.   Length 4½ lines.

Amarygmus suturalis.

A. breviter ovalis, sat fortiter convexus, aterrimus, nitidus; elytris
purpureo-cupreis in fusca mutantibus, sutura viridi, fortiter
striato-punctatis.

Hab. Swan River.

Shortly oval, rather strongly convex, deep glossy black;
the elytra purplish-copper changing to brown, the suture bright

green; head flattish above the clypeus, the latter convex; eyes moderately approximate; antennæ stoutish, especially outwards, the last joint irregularly and broadly ovate; prothorax broad at the apex, rather narrow at the base, minutely punctured; scutellum convex, triangular; elytra striate-punctate, the striæ narrow and rather deep, the punctures small and nearly contiguous, the intervals of the striæ almost impunctate; body beneath and legs glossy black. Length 5 lines.

This and the two above are among the very few striated species of the genus; and of the striated species they are the most convex and elliptical in outline. Besides the difference of colour, *A. suturalis* has the antennæ much stouter and the elytra much more deeply striate than *A. ellipsoides.*

Amarygmus torridus.

A. breviter ovalis, convexus, nitidus; prothorace fulvescenti-cupreo; elytris viridi-metallicis, fortiter seriatim punctatis; corpus subtus femoribusque castaneo-fuscis.

Hab. Cape York.

Shortly oval, convex, shining; head black; clypeus very broad; antennæ reddish brown, slightly thicker outwards, extending to half the length of the body, third joint longest, the rest of nearly equal length; prothorax yellowish copper, closely and finely punctured; scutellum equilaterally triangular, black; elytra about a quarter longer than broad, convex, coarsely seriate-punctate, all the punctures about equidistant from one another; body beneath and femora dark chestnut-brown, slightly nitid; tibiæ and tarsi reddish brown. Length 5½ lines.

In form something like *A. convexus,* but shorter. An isolated species.

EXPLANATION OF THE PLATES.

PLATE X.

Fig. 1. *Melytra ovata:* a, mentum, lower lip, &c.; b, maxilla &c.
Fig. 2. *Blepegenes aruspex:* a, mentum &c.; b, maxilla &c.; c, head.
Fig. 3. *Hymæa succinifera:* a, mentum &c.; b, maxilla &c.
Fig. 4. *Seirotrana crenicollis.*
Fig. 5. *Coripera ocellata.*
Fig. 6. *Byallius reticulatus:* a, mentum &c.; b, maxilla &c.
Fig. 7. *Ganyme Howittii:* a, antennæ.
Fig. 8. *Orcopagia monstrosa:* a, mentum &c.; b, maxilla &c.; c, antenna; d, head; e, fore tibia. N.B. The figure is much too broad in proportion.
Fig. 9. Coxa and part of the femur of a *Pimelia:* a, the trochantin; b, the trochanter. The left side is supposed to be towards the spectator.

PLATE XI.

Fig. 1. *Ectyche erebea:* a, mentum &c.; b, maxilla &c.
Fig. 2. *Meneristes laticollis.*
Fig. 3. *Asphalus ebeninus:* a, maxilla; b, mentum.
Fig. 4. *Tanylypa morio:* a, mentum; b, maxilla; c, fore leg.
Fig. 5. *Brises trachynotoides:* a, mentum; b, maxilla.
Fig. 6. *Ephidonius acuticornis:* a, mentum; b, maxilla.
Fig. 7. *Apasis Howittii* (♂).

PLATE XII.

Fig. 1. *Tyndarisus longitarsus.*
Fig. 2. *Altes binodosus:* 2 a, the same in profile.
Fig. 3. *Mithippia aurita:* 3 a, meso- and metasterna.
Fig. 4. *Helæus squamosus.*
Fig. 5. *Barytipha socialis.*
Fig. 6. *Hectus anthracinus.*

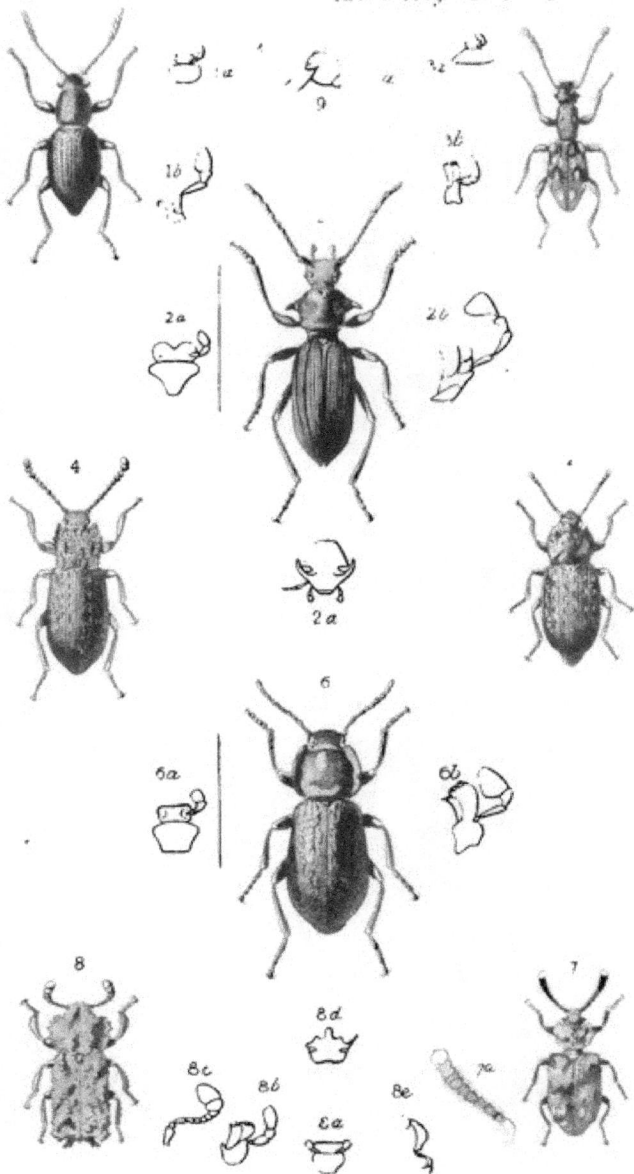

4

4c

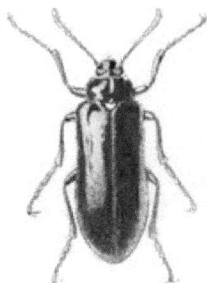

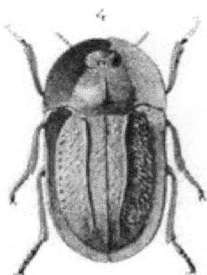

www.ingramcontent.com/pod-product-compliance
Lightning Source LLC
Chambersburg PA
CBHW031410270326
41929CB00010BA/1395

* 9 7 8 3 3 3 7 2 2 1 7 3 7 *